IEA Research for Education

A Series of In-depth Analyses Based on Data
of the International Association for the Evaluation
of Educational Achievement (IEA)

Volume 3

The International Association for the Evaluation of Educational Achievement (IEA) is an independent nongovernmental nonprofit cooperative of national research institutions and governmental research agencies that originated in Hamburg, Germany, in 1958. For nearly 60 years, IEA has developed and conducted high-quality, large-scale comparative studies in education to support countries' efforts to engage in national strategies for educational monitoring and improvement.

IEA continues to promote capacity building and knowledge sharing to foster innovation and quality in education, proudly uniting more than 60 member institutions, with studies conducted in more than 100 countries worldwide.

IEA's comprehensive data provide an unparalleled longitudinal resource for researchers, and this series of in-depth thematic reports can be used to shed light on critical questions concerning educational policies and educational research. The goal is to encourage international dialogue focusing on policy matters and technical evaluation procedures. The resulting debate integrates powerful conceptual frameworks, comprehensive datasets and rigorous analysis, thus enhancing understanding of diverse education systems worldwide.

More information about this series at http://www.springer.com/series/14293

Oliver Stacey · Giulia De Lazzari
Hilary Grayson · Hazel Griffin
Emily Jones · Amanda Taylor
David Thomas

The Globalization of Science Curricula

Oliver Stacey
National Foundation for Educational Research
Slough, Berkshire
UK

Emily Jones
National Foundation for Educational Research
Slough, Berkshire
UK

Giulia De Lazzari
National Foundation for Educational Research
Slough, Berkshire
UK

Amanda Taylor
National Foundation for Educational Research
Slough, Berkshire
UK

Hilary Grayson
National Foundation for Educational Research
Slough, Berkshire
UK

David Thomas
National Foundation for Educational Research
Slough, Berkshire
UK

Hazel Griffin
National Foundation for Educational Research
Slough, Berkshire
UK

ISSN 2366-1631   ISSN 2366-164X (electronic)
IEA Research for Education
ISBN 978-3-319-71531-5   ISBN 978-3-319-71532-2 (eBook)
https://doi.org/10.1007/978-3-319-71532-2

Library of Congress Control Number: 2017960914

This Springer imprint is published by Springer Nature
The registered company is Springer International Publishing AG
The registered company address is: Gewerbestrasse 11, 6330 Cham, Switzerland

Foreword

IEA's mission is to enhance knowledge about education systems worldwide and to provide high-quality data that will support education reform and lead to better teaching and learning in schools. In pursuit of this aim, it conducts, and reports on, major studies of student achievement in literacy, mathematics, science, citizenship, and digital literacy. These studies, most notably TIMSS, PIRLS, ICCS and ICILS, are well established and have set the benchmark for international comparative studies in education.

The studies have generated vast datasets encompassing student achievement, disaggregated in a variety of ways, along with a wealth of contextual information which contains considerable explanatory power. The numerous reports that have emerged from them are a valuable contribution to the corpus of educational research.

Valuable though these detailed reports are, IEA's goal of supporting education reform needs something more: deep understanding of education systems and the many factors that bear on student learning advances through in-depth analysis of the global datasets. IEA has long championed such analysis and facilitates scholars and policymakers in conducting secondary analysis of our datasets. So we provide software such as the International Database Analyzer to encourage the analysis of our datasets, support numerous publications including a peer-reviewed journal— *Large-scale Assessment in Education*—dedicated to the science of large-scale assessments and publishing articles that draw on large-scale assessment databases, and organize a biennial international research conference to nurture exchanges between researchers working with IEA data.

The IEA Research for Education series represents a further effort by IEA to capitalize on our unique datasets, so as to provide powerful information for policymakers and researchers. Each report focuses on a specific topic and is produced by a dedicated team of leading scholars on the theme in question. Teams are selected on the basis of an open call for tenders. The intention is to have two such calls a year. Tenders are subject to a thorough review process, as are the reports produced. (Full details are available on the IEA website.)

The present report is concerned with science education and, specifically, the extent to which science curricula in schools are converging around the world. There are ongoing discussions about the extent to which international large-scale assessments impact on national curricula and lead to curriculum convergence. So far, the underpinning evidence is largely unsystematic and anecdotal. IEA TIMSS data on national curricula, gathered systematically over many years, are excellent resources to investigate what is actually happening and ground these discussions in solid evidence. The goal of this publication was to identify real changes in science curricula and relate these to country contexts.

Conducted by researchers at the National Foundation for Educational Research in England and Wales (NFER) in the United Kingdom, *The Globalization of Science Curricula* draws on the TIMSS encyclopedias and questionnaires to identify changes in science curricula over the past 20 years and elucidate the emergence of elements of a core international science curriculum. Given the internationalization of science and the importance of science education in schools, this report will be a key resource for policymakers, curriculum designers, and teacher educators.

A forthcoming report in this series will deal with teaching tolerance in a globalized world.

Seamus Hegarty
Chair of IEA Publications and Editorial Committee

Contents

Chapter 1
Has Globalization Impacted Science Curricula? An Introduction

Abstract Globalization is a powerful process that exerts an increasing influence on many aspects of society. The impact of globalization on education, and more particularly its impact on the curriculum, is an interesting topic for research, but depends on acquiring comparable data on school curricula from sufficient numbers of countries. The IEA's Trends in Mathematics and Science Study (TIMSS) has collected data on the mathematics and science curricula of participating countries since the 1990s that enables investigation of the national content of science and mathematics curricula over time. Because existing research has tended to focus on mathematics curricula, this study focuses on the evolution of science curricula. TIMSS asks specific questions about the intended curricula, and while the intended curriculum is not necessarily what was implemented or achieved, it has a strong influence on the implemented and achieved curricula of an education system. Many other factors, including local cultural influences, may also contribute; the influence of the international large-scale assessments themselves may lead to countries adopting education reforms and policies that have been successfully implemented by high-performing jurisdictions. Understanding whether and why there have been identifiable global changes resulting in a putative international core curriculum may reveal which strategies and topics countries have recognized as supporting future skills and knowledge.

Keywords Globalization · Science education · Science curriculum
Trends in Mathematics and Science Study (TIMSS)

1.1 Establishing a Framework

Globalization is a topic that is frequently in the news, with the economic and sociopolitical impacts of globalization often receiving a considerable media coverage. By contrast the effects of globalization on education and curricula receive far less attention. Nonetheless, the potential impacts of globalization on education are of international interest, and some have argued that competition between countries

© International Association for the Evaluation of Educational Achievement (IEA) 2018
O. Stacey et al., *The Globalization of Science Curricula*, IEA Research
for Education 3, https://doi.org/10.1007/978-3-319-71532-2_1

for labor force, the mobility of people, and the influence of intergovernmental agencies, such as the Organisation for Economic Cooperation and Development (OECD) or the European Union, are promoting increasing similarities among countries' education systems and curricula. Supranational organizations, such as the World Bank, may also exert their influence on educational systems, particularly in developing countries.

But is education around the world really becoming increasingly similar? Does evidence exist to support the notion that the curricula developed by ministries of education and educational practice in schools are increasingly conforming to a set of "international standards" and, if so, is it possible to identify these benchmarks?

Where research studies on the globalization of curricula do exist, they have tended to focus on mathematics curricula as opposed to science curricula (Rutkowski and Rutkowski 2009; Zanini and Benton 2015). Science education is an important subject to investigate from a global curriculum perspective because science has always been an international discipline, with important discoveries and advances being made all over the world, but with the need for a central body of understanding for these to be interpreted and presented as a school curriculum. Therefore, internationalized thinking has long been an influence on the development of science curricula, arguably more so than in subjects such as literature or history, which are more likely influenced by national cultural and historical perspectives.

Many of the topics taught as science in schools are directly influenced by global factors, such as human health and disease, environmental impacts including climate change, and the future of food and energy supplies. It can also be argued that scientific development is the predominant driver of economic growth and quality of life. Consequently, understanding how science is intended to be taught provides insight into how future citizens are taught about the nature and role of scientific advances.

The IEA's Trends in Mathematics and Science Study (TIMSS) has collected extensive data on intended mathematics and science curricula of participating countries since the 1990s (Mullis et al. 2016). While the intended curriculum is not necessarily what was implemented or achieved, it nonetheless has a strong influence on the implemented and achieved curricula of an education system. Twenty years of collected data provides a valuable opportunity to directly explore the effects of globalization on countries' science curricula over time.

Given the global nature of science and science curricula, this study aimed to address three central research questions:

Have there been changes in intended science curricula over the last 20 years?

To answer this question, we conducted a coding exercise in which changes in the intended science curriculum of countries participating in TIMSS were tracked using responses to the TIMSS curriculum questionnaire. This not only enabled the identification of changes in the curriculum but also enabled the nature of changes to be tracked, for example, if countries were adding new topics to their curriculum or removing topics from the curriculum.

If changes do exist, do they support the hypothesis that science curricula are becoming increasingly similar across countries?

To address this question, we used the outcomes from our coding exercise. We also conducted cluster and discriminant analyses on countries' responses to the TIMSS curriculum questionnaires. This enabled us to identify groups of countries that included similar science content in their curriculum. Finally, for a smaller sample of countries, information on additional features of the science curriculum, such as the mean time spent teaching science per year, was obtained from the TIMSS encyclopedias and TIMSS international results in science reports (for a full list of available TIMSS publications, see https://timssandpirls.bc.edu/isc/publications.html).

Are there groups of countries where curricula are increasingly similar; can the basis of an international core curriculum be identified?

We considered the outcomes of the coding exercise and cluster and discriminant analyses jointly in order to establish whether the intended science curricula of TIMSS countries were becoming increasingly similar. Interpreting the results and outcomes of both approaches enabled us to establish a more complete picture of any emergent international core curriculum.

1.2 Overview of This Book

After this brief introduction, a literature review (Chap. 2) provides background information and context. The literature review has three key aims (1) to identify the factors contributing to the globalization of science education and science curricula; (2) to identify existing evidence for the globalization of science curricula over time; and (3) to identify the methods that have been used previously to investigate the globalization of curricula.

The review establishes potential drivers and mediators for the globalization of science curricula. Different research approaches are identified, ranging from qualitative methods such as interviews, lesson observations and analysis of curriculum documents, through to quantitative methods using statistical approaches such as cluster analysis and latent class analysis.

This literature review informs the different methods we use to investigate the research questions (Chap. 3). Using multiple methodological approaches enables evidence to be collected on different aspects of our three central research questions. Countries participating in TIMSS show considerable variation in terms of the number and nature of changes they made to their science curriculum between 1999 (2003 for Grade 4) and 2015 (Chap. 4), and using this we are able to identify and highlight topics that were included in the vast majority of participating countries' intended science curricula. Cluster analysis and discriminant analysis provide evidence for convergence or divergence in science curricula over time. In Chap. 5, we examine how the evidence obtained from the analyses and the literature review relate to our three research questions.

Our research contributes to the understanding of globalization in science education and curricula, however, this study is not a measure of the total science curriculum for each country considered, but a measure of the similarity of their intended science curriculum to the TIMSS framework. This research only examined the science curricula of countries that participated in TIMSS, but the same approach could be applied to other international large-scale assessments to assess their impact on national science curricula. We conclude in Chap. 6 by suggesting future areas of research which would complement this study.

References

Mullis, I. V. S., Martin, M. O., & Loveless, T. (2016). *20 Years of TIMSS: International trends in mathematics and science achievement, curriculum, and instruction.* Chestnut Hill, MA: TIMSS & PIRLS International Study Center, Boston College. Retrieved from http://timss2015.org/timss2015/wp-content/uploads/2016/T15-20-years-of-TIMSS.pdf.

Rutkowski, L., & Rutkowski, D. (2009). Trends in TIMSS responses over time: Evidence of global forces in education? *Educational Research and Evaluation, 15*(2), 137–152.

Zanini, N., & Benton, T. (2015). The roles of teaching styles and curriculum in mathematics achievement: Analysis of TIMSS 2011. *Research Matters, 20,* 35–44.

Chapter 2
A Review of the Existing Literature on Globalization of Science Curricula

Abstract Given the extensive nature of globalization and its wide-ranging impact, the review of the literature on the effect of globalization on science education and science curricula was restricted to three particular aspects. Firstly, identifying the factors that potentially contribute to the globalization of science education; secondly, exploring research evidence for the globalization of science curricula over time; and thirdly, analyzing the approaches and methods that have been used in previous research studies to empirically investigate the globalization of curricula. A wide range of factors were identified as promoting the globalization of science education and curricula, including the growing use of information technologies and the increasing influence of intergovernmental organizations in education. However, regional and local cultural factors may counteract globalization to some extent. The literature review identified some evidence for globalization of science curricula over time; the impacts of international large-scale assessments, such as the Trends in International Mathematics and Science Study (TIMSS), may have driven changes and reforms in the science curricula of many participating countries. The literature revealed that a number of both qualitative and quantitative methods had been used to investigate globalization of curricula over time, and the advantages and drawbacks of each of the approaches were considered.

Keywords Curriculum alignment · Curriculum convergence · Globalization International large-scale assessment · Science education · Science curriculum Trends in Mathematics and Science Study (TIMSS) · TIMSS video study

2.1 Introduction

Globalization is a process with far-reaching impacts in many different disciplines, including education. Globalization is an extensive concept, although in its simplest form it can be described as relating to "reforms and structures that transcend national borders" (Astiz et al. 2002). Within education, the process of globalization has the potential to exert a wide range of effects on national education policies and school science curricula. Whilst globalization has distinctive effects on education, it is important to note that this cannot be completely removed from wider economic and cultural globalization.

Given the variety of ways in which globalization can impact education and curricula, this literature review had three key aims:

(1) to identify and explore the factors contributing to globalization in science education and science curricula
(2) to investigate the research evidence for the globalization of science curricula over time
(3) to identify the approaches and methods that have been used in previous research studies to empirically investigate the globalization of curricula, and to use these to inform the statistical analyses conducted during this research project.

A literature search was conducted during the autumn of 2016. Given the large number of published articles on globalization in education and the resources available for this report, we did not attempt to present an exhaustive overview of all published articles relating to globalization of curricula. Instead, we employed a highly focused search strategy. The literature search was designed to identify studies that provided evidence of countries that had redesigned or modified their science curricula to mirror high-performing jurisdictions or to align more closely to the TIMSS framework (curriculum) content. The literature search also looked for studies that had already employed TIMSS data to investigate globalization of curricula.

The search strategy was designed to maximize the chances of identifying the most pertinent literature within those constraints. The date parameters were limited to recent (2010 onwards) evidence on issues surrounding global alignment/ globalization of science curricula. However, the date parameters were extended to 1995 to identify research that had used TIMSS data to explore issues surrounding global alignment/globalization of science curricula (the influential Third International Mathematics and Science Study, the first cycle of what later became known as TIMSS, was administered in 1995). This ensured that significant studies based on data from the 1995 study were not missed from the literature review.

Key education bibliographic databases (the Australian Education Index, the British Education Index and the Education Resources Information Center [ERIC]) were searched using globalization/curriculum alignment terms combined with science education terms tailored to the specific search capability of each database (see Appendix A for a detailed explanation of the search terms employed). In addition, the table of contents of the key science education journal, the *International Journal of Science Education*, were reviewed from 2010 onwards and relevant research papers were noted. The publications list on the International Association for the Evaluation of Educational Achievement (IEA) website was also reviewed.

The literature search identified 76 research articles. The abstracts of each of these articles were then appraised by three researchers in order to identify the most relevant articles to include in the literature review. A number of quality criteria were applied to establish the relevance of the article to the literature review. Priority was given to: articles that considered the curricula/education system of more than one country; articles published in peer-reviewed journals; articles that focused on science curricula (as opposed to other subject curricula); empirical studies; and studies that made use of data from international large-scale assessments such as TIMSS. A set of questions was used to appraise the articles identified in the literature search (Appendix B).

We used a template (see Appendix C) to review and evaluate the contents of the ten most relevant articles and summarize any data or evidence related to our three research questions.

The research in this literature review identified forces that are contributing to and driving globalization in science education and science curriculum and also forces that are counteracting or mediating globalization.

2.2 Factors Contributing to the Globalization of Science Education and Science Curricula

Spring (2008) identified a number of different forces that are involved in driving globalization within education more generally. These forces include:

- the emphasis on education as a mechanism for economic growth
- intergovernmental, governmental and international non-governmental organizations
- information technology
- multinational corporations
- international large-scale assessments.

Whilst these forces are shaping globalization in education in a more general sense, they are also influencing and impacting the globalization of science education and science curricula. The specific impact of each of these forces is outlined below. Although each force is discussed separately, it is important to appreciate that these forces do not exist in isolation and interact with each other to shape the process of globalization in science education.

2.2.1 Education as a Mechanism for Economic Growth

The twenty-first century has seen growing convergence in political agendas in relation to education, particularly in the developed world. Spring (2008) identified the concept of the *knowledge economy* as one of the most significant factors contributing to the globalization of education and curricula. As countries around the world become more engaged in the globalized economy, there develops an important need for nations to prepare their young people for active participation within these globalized markets (Sellar and Lingard 2014). Also, as countries seek to build and develop knowledge-based economies, this has a direct effect on education and school curricula as politicians seek to equip learners with the skills and competencies that are needed to succeed in the global economy. Weber (2011) identified this phenomenon in the Gulf states, where the focus on the knowledge economy is influencing school curricula as politicians seek to diversify these states from their current hydrocarbon-based economies.

The knowledge economy is linked to the increased focus in many countries on the role of education in promoting economic growth. As a consequence, there has been greater emphasis in many countries on schools developing and equipping students with workplace skills and competencies and preparing them for life beyond education. This in turn has the potential to lead to convergence in science curricula as countries adapt curricula to focus on aspects of science that have more potential to facilitate future economic growth.

Globalization and the economic aspects of globalization are of particular relevance to science education and science curricula, as many of the most globalized sectors of the economy have a technological or scientific basis that requires specific scientific knowledge and skills. Computer and mobile technologies, pharmaceuticals and biotechnologies, petrochemicals and emerging clean energy technologies are for example some of the most globalized sectors of the world economy. Therefore, countries seeking to develop economic engagement in these areas may consider changes to science curricula as a medium- to long-term strategy for successful involvement in these sectors.

2.2.2   Intergovernmental and Non-governmental Organizations

In addition to the knowledge economy and economic globalization, some intergovernmental and nongovernmental organizations are important actors in the globalization of education. For example, intergovernmental organizations such as the World Bank and the Organisation for Economic Cooperation and Development (OECD) are influential in both shaping educational discourse and the educational agenda in many countries. Both organizations view the purpose of education from an economic perspective and consider education as a mechanism from which to stimulate economic growth. Sellar and Lingard (2014), for example, suggested that "the rise of the OECD's education work is linked to the economization of education policy". These organizations consider one of the primary roles of school is to prepare students to be successful participants in the knowledge economy (OECD 1996).

2.2.3 Information Technology and Multinational Corporations

The increasing role of information technology and the internet has had a significant impact on globalization in science education. This has largely been due to the speed and ease with which information can be accessed and transferred (OECD 1996). Potential impacts of this on science curricula include the rapid sharing of scientific information and ideas across borders by universities and educational institutions, as well as multinational corporations who provide educational services and curriculum resources to schools and education ministries across the world.

Information and communications technology (ICT) has already transformed science education and curriculum in a number of different ways, for example by "expanding the spaces, methods and times for its spread" (Cornali and Tirocchi 2012). Science teaching materials have been enhanced by ICT, with added functionalities including animations and multimedia content. In science education, this has meant that experiments that could not be undertaken in the average school science laboratory may now be simulated by such teaching tools. The quantity and accessibility of learning resources has increased dramatically with the proliferation of ICT, and personalized learning tailored to the needs of the individual student has become more feasible and cost effective. ICT has also allowed instantaneous sharing of information and content and permitted interactions between learners over larger geographic distances. In future, the impact of ICT on science education and curricula is likely to increase further, as traditional science curricula based on factual knowledge are replaced with more open science curricula centered on the acquisition of specific skills (Cornali and Tirocchi 2012).

2.2.4 International Large-Scale Assessments

One of the most powerful forces shaping globalization in science education and science curricula is the development of international large-scale assessments of science such as TIMSS and the Programme for International Student Assessment (PISA). The PISA assessment has been described as playing "a major role in the standardization of education" (Spring 2008). International assessments are growing in influence as the number of countries participating in them increases. Large-scale assessment is contributing to the globalization of science education and science curricula in a number of different ways, as we will discuss.

A review of the effects of TIMSS 1995 on teaching and learning in 29 different countries highlighted the role that it had played in the restructuring of national science curricula and the subsequent integration of science curricula across some countries (Robitaille et al. 2000). Following TIMSS 1995, a number of the participating countries embarked on extensive science curriculum revisions, changing

Table 2.1 Responses to TIMSS 1995 science results

Type of response	Description of response	Example countries
No major changes to curriculum	No changes or very minor changes in response to TIMSS	Japan, Netherlands, Flemish Belgium
Changes to content in science curriculum	Extra topics added to the science curriculum, for example the inclusion of environmental science topics	Iran, Kuwait
Changes to skills emphasis in curriculum	Shift from acquiring knowledge to being able to apply knowledge; greater emphasis on practical and problem-solving skills	Latvia, Czech Republic
Increased status of science within education system	More time allocated to teaching science, particularly at primary age	Iceland
Change to structure/ organization of science curriculum	Introduction of an integrated science curriculum in the primary phase of schooling	Romania
Changes to science assessment	Change from sampling assessments to full cohort testing	Philippines
	Amendments to science assessments to incorporate TIMSS items	British Columbia (Canada)
Changes to teacher training and development	Increased support and training to strengthen pre-secondary school teachers' science knowledge	Norway
Changes to curriculum to address student attitudes	Science curriculum amended to address students' negative attitudes towards science	Republic of Korea

the science content taught (as in Kuwait) or the skill areas emphasized within the curriculum (as in Latvia) (see Table 2.1).

Evidence from Israel (Klieger 2015) supports the idea that international surveys and the content assessed in them can promote the convergence and globalization of science curricula. Following poor TIMSS results in both 2003 and 2007, the Israeli Ministry of Education opted to reform its science curriculum so that it was more aligned to the TIMSS science content domains. Therefore, in some countries at least, the TIMSS assessments have directly affected national science curricula.

By contrast, interviews with experts in science education in Australia (a country which has traditionally performed relatively well in TIMSS and PISA) suggest that the international assessments have not had a major impact on the curriculum (Aubusson 2011). However, these interviews did suggest that TIMSS and PISA were nonetheless influential, and there was a clear desire amongst Australian policymakers and the science education community to ensure that standards on these assessments were maintained over time.

Robitaille et al. (2000) also noted that for some countries (such as Japan), the impact of TIMSS on the science curriculum had been relatively minor. Differences in the impact that international large-scale assessments have had on countries' curricula reveal that each country participating has "its own unique set of motivations for participating, and each of them has their own set of expectations for the study" (Robitaille et al. 2000).

2.2.5   Policy Borrowing and Standardization

Linked to the rising influence of international large-scale assessments is "policy borrowing", where countries adopt education reforms and policies that have been successfully implemented in other countries, typically high-performing jurisdictions. There has been a suggestion that the rate of policy borrowing is increasing over time, and that this is directly affecting national education policies (Rutkowski and Rutkowski 2009). This increase in policy borrowing is perhaps unsurprising, as international assessments facilitate comparison between countries and provide evidence for the most successful and high-performing countries (as measured by the assessment).

In addition to policy borrowing and international large-scale assessments, Astiz et al. (2002) outlined the global trend towards greater emphasis on standardization, achievement and assessment in science and mathematics curricula. This has been achieved through approaches strengthening school accountability and is also contributing to greater convergence of curriculum goals in different countries.

However, despite the wide range of forces driving globalization in science education and curricula, there are other factors that are counteracting or mediating the rate of globalization in this area. One of the most important counteracting factors is local culture. For example, research comparing science teaching in Grade 6 classrooms in Australia and China highlighted the impact of culture on curricula and educational practices in these countries (Tao et al. 2013).

The study found that the contrasting cultures and education philosophies in Australia and China had an impact not only on classroom practices but also on how education reforms are enacted. For example, in China, despite reforms to the science curriculum to give a greater emphasis to constructivist approaches, the research identified resistance to these reforms, with traditional teaching approaches (such as the memorization of facts, reading books and watching teachers conduct experiments) still the dominant approach to science teaching in many schools. Clearly, even if the intended science curricula in different countries or jurisdictions may be becoming more globalized, this does not necessarily mean that the implemented science curriculum experienced by students becomes more closely aligned.

The misalignment between countries' intended and implemented science curricula and the complex interplay between central and local forces is not limited to China. Astiz et al. (2002) identified similar issues in Spain and the United States. In Spain, a common curriculum is specified by central government, but this curriculum is then adapted and interpreted locally in the different regions of the country.

This means that students in the different regions will have different interpretations of the same curriculum.

In the United States, there are several contrasting forces that exert an influence on the curriculum. On one level there has been the centralization of curriculum goals, but opposing this there has been decentralized curriculum implementation. Once again, the localization of curriculum interpretation and implementation means that the implemented curriculum is likely to be far less standardized and more variable than the intended curriculum prescribed centrally. These countries provide strong examples of how global forces act on curriculum at a national level but are then modified and adapted at a more local level.

Other evidence also suggests that the effect of globalization in science education and science curricula is relatively limited. Research using TIMSS data, which compared the content standards, textbook content and teaching time allocated to science topics for students at Grade 8 in participating countries, indicated that science teaching and science curricula in Grade 8 in different countries were far less homogenous than for mathematics (Cogan et al. 2001). Considerable differences in countries' approaches to science education have also been attributed to cultural differences between countries: "how the curriculum is specified and organized, what students are expected to learn and be able to do – are all reflections of that culture" (Cogan et al. 2001, p. 106).

This preliminary overview reveals a number of different factors have been identified as contributing to the globalization of science education and science curricula. Whilst there are strong driving forces promoting globalization of science education, evidence on the extent of their impact and for convergence in science curricula is variable, with no consistent pattern to the effect on countries. The next section of the literature review specifically considers the evidence for the globalization of science curricula over time.

2.3   What Evidence is There for the Globalization of Science Curricula Over Time?

As noted in Sect. 2.2.4, each country has its own set of local conditions which impact both curriculum decisions and their motivation for participating in international assessments such as TIMSS and PISA. However, several of the research studies identified in our literature review provided evidence for the globalization of science curricula over time, identifying international large-scale assessments such as TIMSS and PISA as playing a role in this convergence. This is, in part, due to the way that these assessments facilitate the comparison of different education systems. This, in turn, may act as a catalyst for change and convergence, with countries making changes to their education systems in order to address perceived weaknesses or deficiencies identified as a result of their participation in the international assessments.

The TIMSS 1995 survey provides a strong early example of the impacts of international large-scale assessments on science education and curricula, with many governments analyzing the results from TIMSS 1995 and using the outcomes to inform decisions about future educational and economic development.

2.3.1   How Different Countries Responded to TIMSS 1995

Robitaille et al. (2000) analyzed the changes that were made to science curricula in a wide range of countries following the publication of the TIMSS 1995 results. Countries' responses to their performance in TIMSS could be sorted into several broad categories (Table 2.1).

For some high-performing countries, such as Japan, the TIMSS 1995 results had minimal impact on their science curriculum. Other countries, which performed reasonably well in TIMSS 1995, used the wide range of data reported in the study to modify specific aspects of their science curriculum. For example, in the Republic of Korea, the government reduced the overall content of the curriculum to address students' negative attitudes to science, as reported in the TIMSS data. Both the Czech Republic and Latvia modified performance expectations to promote the application of science and problem-solving skills.

Kuwait introduced a module on environmental science and Romania reorganized its curriculum into a more coherent whole to improve students' abilities to make connections between different curriculum areas. Some changes were more closely related to policy or application of policy than the specific curriculum. For example, Norway placed a much greater emphasis on strengthening the subject knowledge of teachers, whilst the Canadian province of British Columbia introduced new assessments for Grades 4 and 7, with about half of the assessment questions drawn from released TIMSS items.

For several countries, a relatively poor performance in TIMSS 1995 acted as a direct prompt to institute significant science curriculum change, in an attempt to improve their performance in future international comparison tests. In Iceland poor performance in TIMSS 1995 started a process of curriculum review where the TIMSS framework itself was used as a curriculum model, leading to an increase in the importance of science and mathematics teaching in Icelandic schools. Similarly, in Iran, poor results in TIMSS 1995 led to the identification of factors that merited increased attention in the curriculum, prompting changes to the curriculum again based on the TIMSS framework. These increased the focus on scientific skills, as well as the cognitive demand of the curriculum, and extended curriculum coverage to include areas such as the environment.

2.3.2 Case Study: Israel

Israel provides an interesting case study of how a country has responded specifically to the results from international large-scale assessments by changing its curriculum and how this was implemented. Klieger (2015) outlined a series of policy changes made between 1996 and 2011; the country's poor performance in the international surveys provided a major incentive for reforms, leading, in 2009, to the Ministry of Education setting achievement targets to progress at least ten places in the rankings for the TIMSS 2011 and PISA 2012 assessments. They aimed to achieve this by intervening at both the intended and implemented curricular levels, through a series of curriculum changes and operational changes to the delivery of the curriculum. Klieger (2015) reviewed the evidence for those changes from government documents and compared the science curriculum from 1996 to that of 2011 for evidence of globalizing effects. International surveys had clear impact on the content of the Israeli science curriculum. Topics that were absent from the international frameworks were dropped, including "the Earth and the universe" (some sub-topics were integrated within other topics), "information and communication" and "the senses", whilst topics that appeared in the international frameworks were added, including "human health" and "acids and bases". The structure of the curriculum and the way that skills were presented and organized within the curriculum were also aligned to the international frameworks. For example, the 2011 Israeli curriculum integrated problem-solving skills within the science content and further emphasized high-order thinking skills (HOTS) such as argumentation. In line with international frameworks, scientific inquiry skills continued to be presented separately.

The Israeli Government also sought to address the implemented curriculum by interventions at the teacher level, seeing this as a more immediate way to bring about change than through the intended curriculum. For example, in 2005, advice was issued to ensure that teachers taught the content of international surveys even when it did not appear in the Israeli curriculum at that time. In 2009, there was an intensive in-service training program for teachers based on the findings of the TIMSS and PISA studies, accompanied by closely specified "kits" of teaching materials that teachers were obligated to use.

The desired impact of these changes to the curriculum and its implementation was seen in the results from TIMSS 2011. The baseline achievement for TIMSS 1995 (Grade 8) in Israel was an average scaled score of 486, placing the country 26th out of the 38 participating jurisdictions. After the implementation of many of the changes highlighted above, the average scaled score for Israel had risen to 516 by 2011. This exceeded the scaled international average of 500 and ranked the country 13th out of the 42 participating educational jurisdictions. Klieger (2015) concluded that, because of the policies of the Department of Education, the international frameworks have had a conspicuous influence on the curriculum content, but, as a consequence, classroom activities have become much more prescriptive over this time.

2.3.3   Factors That Oppose the Globalization of Science Curricula

The influence of globalizing factors on the intended curricula, through policy-makers engaging with globalizing influences such as international large-scale assessments, has been documented (Klieger 2015; Robitaille et al. 2000). However, the implemented curriculum is delivered by teachers in the classroom, who are exposed to a different set of influences, especially at a more local, cultural level. Cogan et al. (2001) examined how different localized educational "cultures" can affect the implementation of curricula by analyzing the TIMSS 1995 dataset to look for culturally specific patterns in the teaching of science across 36 of the countries that participated. In this context, the educational culture is defined in terms of

> how schooling is organized – the goals and purposes identified for each year, how decision making authority is distributed (or not)... how the curriculum is specified and organized, what students are expected to learn and be able to do.

(Cogan et al. 2001, p. 106).

Cogan et al. (2001) analyzed the different curricula at three levels: the intended curriculum, the potentially implemented curriculum, and the implemented curriculum. Firstly, the intended curriculum was measured by coding the different content statements specified in the curricula of the educational jurisdictions. The study compared these to the 48 TIMSS science framework topics for Grade 8 assessment. Whilst most countries specified teaching the majority of these topics, there were significant variations. Seven countries included fewer than half of these topics in their curricula (Republic of Korea, Hong Kong, Romania, Japan, Germany, Greece and the Czech Republic), whilst three countries included all topics (Iran, New Zealand and the United States).

The study quantified the potentially implemented curriculum by coding the content of science textbooks. Whilst most countries' textbooks covered the majority of the 48 TIMSS science framework topics, there were six countries that covered less than half of these in textbooks (Denmark, Japan, Iran, Singapore, Israel and Germany), whilst four (Canada, Colombia, Switzerland and the United States) covered all, or nearly all topics.

The implemented curriculum was measured by the allocation of teachers and teaching time to the different topics. The picture in terms of teaching time is complicated by the teaching of science as different courses (e.g. biology, chemistry, physics and earth sciences), which means that no single topic in science was taught by more than 70% of all science teachers (compared to an average of 90% of all teachers for mathematics topics). This produced a disparate picture, with substantial variations apparent from a statistical analysis of the percentage of teachers teaching specific topics in different countries.

In conclusion, Cogan et al. (2001) indicated that there were few commonalities across countries, or even within countries at different curricular levels. For example, they were unable to define five "core" science topics across participating countries as they derived a different list of topics at each curricular level (curricular content, textbooks, teaching content and instructional time allocated). This variation is in contrast to a similar study conducted for mathematics that did exhibit consistency across countries and levels of organization. They concluded that this "reflects the greater diversity in the way science is organized and delivered across the TIMSS countries, at least in comparison to mathematics" (Cogen et al. 2001, p. 128).

Other studies have reported regional cultural effects. Kjaernsli and Lie (2008) adopted a cluster analysis approach using item responses to TIMSS 2003 science questions and identified a number of countries that clustered together largely along geographic or linguistic lines. For example, the study identified an Arabic cluster, an English-speaking cluster and a South Eastern European cluster. This study provides some evidence for the alignment and convergence of science curricula along geographic and cultural lines.

A more detailed approach to looking at cultural differences in implemented national curricula was taken by the 1999 TIMSS video study (Roth et al. 2006). The study was conducted in conjunction with the IEA by the US National Center for Education Statistics and the US Department of Education under a contract with LessonLab, INC of Los Angeles (for further details see http://www.timssvideo.com/timss-video-study). This study compared science teaching and learning in the United States to that of four higher achieving countries (namely Australia, Japan, the Czech Republic and the Netherlands) by coding the science content and instructional approaches evidenced by video recordings of 439 representative Grade 8 lessons across the five countries.

The study findings showed that science teaching in the United States exposed students to a wide variety of both pedagogical approaches and content, whilst the other countries reflected a common content-focused instructional approach (Roth et al. 2006). However, even within this common approach, there were significant variations in the learning cultures of the different countries, with the Czech Republic focused on whole class discussion, Australia and Japan on connecting ideas through data and inquiry, and the Netherlands employing independent textbook-centered reading and writing activities. The cultural differences meant that each of the countries had a distinct approach to science teaching, providing students with different opportunities to learn science and different visions of what it meant to understand science.

2.3.4 Conclusions

The studies explored in Sect. 2.3 present evidence for the differing global, regional and cultural influences on science curricula. Taken as a whole, these studies imply that whilst there is pressure on policymakers to globalize their intended curricula, there are also local cultural pressures working at the level of implementation and realization that may provide some resistance to this globalizing effect.

It is important to note that our literature review was limited to research on the globalization of science curricula from 2010, with studies making use of TIMSS data extending back to 1995. However, it has been suggested that 1995 may be too late a starting point to detect changes in the globalization of curricula because by this time globalization had already exerted a large influence on science curricula (Rutkowski and Rutkowski 2009). This needs to be taken into account when considering the evidence for globalization in science curricula.

2.4   What Methods Have Been Used Previously to Investigate Globalization of Curricula?

The literature review identified a number of different methods that have been used to investigate globalization of curricula. These methods were evaluated to review their appropriateness and robustness for different contexts and to inform the design of the methodology for this study. The methods used include both qualitative and quantitative approaches (Table 2.2).

Qualitative methods used to investigate the globalization of curricula include the documentary analysis of policy documents. For example, Klieger (2015) examined the effect of international surveys on science education and the science curriculum in Israel between 1996 and 2011 by analyzing curriculum and policy documents and how these changed over the period. The study also compared these documents to the requirements of the international surveys such as the TIMSS content domain.

Table 2.2 Methods used to investigate globalization of curricula

Study	Method	Data source
Klieger (2015)	Documentary analysis of education policy documents and curriculum documents	Israeli Ministry of Education policy and curriculum documents
Tao et al. (2013)	Comparison of teaching and learning of science in Chinese and Australian Grade 6 classrooms	Lesson observations, teacher interviews, student questionnaires
Roth et al. (2006)	Comparison of science teaching in the USA and four high-performing jurisdictions	TIMSS 1999 video study
Cogan et al. (2001)	Median polish analysis	TIMSS 1995 curriculum and teacher questionnaires
Rutkowski and Rutkowski (2009)	Hierarchical cluster analysis and non-linear principal component analysis	TIMSS 1995 and 2003 (Grade 8 mathematics item response data)
Kjaernsli and Lie (2008)	Hierarchical cluster analysis	TIMSS 2003 (science item response data)
Zanini and Benton (2015)	Latent class analysis	TIMSS 2011 Grade 8 mathematics teacher questionnaire

This qualitative approach provided evidence that, over time, the Israeli science curriculum was converging on the requirements of international surveys such as TIMSS.

Another study which contained a significant qualitative element to investigate globalization compared the approaches used to teach science in Australian and Chinese elementary schools (Tao et al. 2013). This study used a multiple comparative case study approach in which three Australian schools and three Chinese schools were paired together for comparison based upon socioeconomic status (high, medium or low). The research consisted of analysis of curriculum documents in each school, lesson observations, school tours, teacher interviews and student questionnaires. The methods used in this research enabled comparison of both the intended and implemented curriculum in Australian and Chinese schools, and observation of the impact of culture on the curriculum in each school.

Other research methods used in the literature review studies identified were predominantly quantitative in nature. One study compared aspects of the implemented science curricula in the United States, Australia, the Czech Republic, Japan and the Netherlands by reviewing eighth grade science lessons from the TIMSS video study (Roth et al. 2006). The content of the science lessons was coded and the commonalities and differences between the content of the lessons in the United States and the four comparator countries were identified. This study concluded that each country had a distinct approach to science teaching, providing students with different opportunities to learn science and different visions of what it means to understand science. Additionally, the science lessons in each country varied in terms of their organizational features and in the extent to which students were actively involved in their science lessons.

Some studies made use of the responses to the TIMSS teacher and curriculum questionnaires to investigate globalization of curricula. For example, Cogan et al. (2001) used TIMSS Grade 8 questionnaire responses to investigate similarities in countries' science curricula. Detailed analysis of the questionnaire responses enabled information about science content standards, textbook content and teaching time for each science topic to be collated for each country. Matrices of the curriculum for each country were produced and median polish analysis conducted in order to look for the effects of country by topic interactions.

Several studies made use of hierarchical cluster analysis to investigate the similarities and differences in curricula. This has been done for mathematics curricula (Rutkowski and Rutkowski 2009) and science curricula (Kjaernsli and Lie 2008). In mathematics, Rutkowski and Rutkowski (2009) investigated 16 countries that participated in TIMSS Grade 8 mathematics in both 1995 and 2003. They conducted an item analysis to establish the strengths and weaknesses in student response patterns in each country and how this compares across countries. A hierarchical cluster analysis was then conducted for the 1995, 1999 and 2003 TIMSS cycles to investigate the degree of alignment in curricula across different countries and how this changed between the three TIMSS cycles.

Kjaernsli and Lie (2008) used a similar approach with TIMSS 2003 science data. Probablity value (p-value) residuals were calculated, which measured how much better or worse students in each country performed on a particular science item compared to what would be expected based on the average student achievement of the country and the overall difficulty of the item. A hierarchical cluster analysis was then performed on each country to investigate how participating countries cluster together based on similarities in the strengths and weaknesses of their students' responses to different types of answer.

In addition to hierarchical cluster analysis, latent class analysis has been used to identify groups of countries with similar curricula. For example, Zanini and Benton (2015) investigated groupings of countries with similar mathematics curricula by conducting a latent class analysis using responses to TIMSS 2011 Grade 8 teacher questionnaires. In this study, teacher responses to questions regarding which mathematics topics were taught to their students were used in the latent class analysis. The analysis found five distinct groupings of countries based on the teacher responses to the questionnaire, thereby providing evidence to suggest some degree of alignment and harmonization in mathematics curricula.

The literature review suggests that there are a range of different methods that can be successfully employed to investigate globalization of curricula. Each of these approaches has its own unique set of advantages and limitations. Among the qualitative studies, the documentary analysis of education policy and curriculum documents has a number of strengths. This approach enables a country's curriculum changes to be scrutinized in depth. The curriculum documents and associated government policy papers also often provide information about the explicit aims of any changes or reforms to the curriculum and the rationale for them. This has the advantage over some other approaches in that it can provide an insight into why changes have been made to the curriculum and what the intended aims or aspirations of the changes were.

There are also disadvantages. Curriculum and policy documents can be time consuming to analyze and review, and this places a natural limit on the number of countries that can be examined given the constraints of time and budget. This type of qualitative approach is thus better suited to studies that consider one or a small number of countries as opposed to a large number of countries. Other disadvantages include the difficulty of obtaining documents in a common language for some countries, as well as the variation in the quality and extent of curriculum and policy documents in different countries; this may make direct comparison between countries challenging. Furthermore, these documents often focus on the intended curriculum as opposed to the implemented curriculum, and so caution needs to be exercised when making inferences about a country's implemented curriculum.

Other qualitative approaches identified in the literature review, such as lesson observations and interviews with teachers and students, have a number of advantages when studying the possible globalization of science curricula. As well as providing the opportunity to observe the implemented curriculum in schools as opposed to the intended curriculum, such studies often take the viewpoints of several different actors in the school system (teachers and students) into account. However, this type of study can be time consuming and expensive, constraints that may influence the number of countries that can be considered. It is also difficult to ensure that the data collected is truly representative when relying on a small sample of observations and interviews.

Analyzing extracts from the TIMSS video study has a number of benefits, including the ability to observe science lessons in different countries and to code for different pedagogic features in those countries. This approach also permits teaching methods in different countries to be compared, and so enables aspects of the implemented science curriculum to be explored. However, the TIMSS video study was only undertaken at one time point for science, and so it is thus difficult to use the data to investigate the globalization of science curricula over time or to estimate how representative the recorded lessons were of typical everyday science teaching. Indeed, it must be acknowledged that teacher and pupil awareness that the lessons were being recorded may have affected the quality or nature of the lessons in some way.

The quantitative approaches identified in the literature review, such as cluster analysis and latent class analysis, have a number of advantages but also some disadvantages over the qualitative approaches outlined. Firstly, the quantitative approaches make it easier to include a greater number of countries in the investigation. This is an important consideration when analyzing the globalization of science curricula across the full cohort of countries taking part in an international survey such as TIMSS. These techniques have the potential to allow for more robust large-scale investigations into the globalization of science curricula over time than the qualitative approaches. Both cluster analysis and latent class analysis are well suited to this type of analysis, and cluster analysis has the additional benefit of being able to be conducted on a smaller dataset than latent class analysis.

Having considered the relative merits of each of the techniques identified in the literature review, and the datasets and resources available for this research, the methods we have chosen to use and our rationale for using them are outlined in Chap. 3 of the report.

References

Astiz, M., Wisemand, A., & Baker, D. (2002). Slouching towards decentralization: Consequences of globalization for curricular control in national education systems. *Comparative Education Review, 46*(1), 66–88.

Aubusson, P. (2011). An Australian science curriculum: Competition, advances and retreats. *Australian Journal of Education, 55*(3), 229–244.

Cogan, S., Wang, H., & Schmidt, W. (2001). Culturally specific patterns in the conceptualization of the school science curriculum: Insights from TIMSS. *Studies in Science Education, 36,* 105–133.

Cornali, F., & Tirocchi, S. (2012). Globalization, education, information and communication technologies: What relationships and reciprocal influences? *Procedia—Social and Behavioral Sciences, 47,* 2060–2069. Retrieved from http://www.sciencedirect.com/science/article/pii/S1877042812026857.

Kjaernsli, M., & Lie, S. (2008). Country profiles of scientific competencies in TIMSS 2003. *Education Research and Evaluation, 14,* 73–85.

Klieger, A. (2015). Between two science curricula: The influence of international surveys on the Israeli science curriculum. *The Curriculum Journal, 26*(3), 404–424.

OECD. (1996). *The knowledge-based economy*. Paris: Organisation for Economic Cooperation and Development. Retrieved from https://www.oecd.org/sti/sci-tech/1913021.pdf.

Roth, K. J., Druker, S. L., Garnier, H. E., Lemmens, M., Chen, C., Kawanaka, T., Rasmussen, D., Trubacova, S., Okamoto, Y., Gonzales, P., Stigler, J., & Gallimore, R. (2006). *Highlights from the TIMSS 1999 video study of eighth-grade science teaching* (NCES 2006-17). US Department of Education, National Center for Education Statistics. Washington, DC: US Government Printing Office. Retrieved from https://nces.ed.gov/pubs2006/2006017.pdf.

Robitaille, R., Beaton, A., & Plomp, T. (Eds.). (2000). *The impact of TIMSS on the teaching and learning of mathematics and science*. Vancouver: Pacific Educational Press.

Rutkowski, L., & Rutkowski, D. (2009). Trends in TIMSS responses over time: Evidence of global forces in education? *Educational Research and Evaluation, 15*(2), 137–152.

Seller, S., & Lingard, B. (2014). The OECD and the expansion of PISA: New global modes of governance in education. *British Educational Research Journal, 40* doi: https://doi.org/10.1002/berj.3120. Retrieved from https://www.researchgate.net/publication/259535939_The_OECD_and_the_expansion_of_PISA_New_global_modes_of_governance_in_education.

Spring, J. (2008). Research on globalization and education. *Review of Educational Research, 78*(2), 330–363.

Tao, Y., Oliver, M., & Venville, G. (2013). A comparison of approaches to the teaching and learning of science in Chinese and Australian elementary classrooms: Cultural and socioeconomic complexities. *Journal of Research in Science Teaching, 50*(1), 33–61.

Weber, A. S. (2011). The role of education in knowledge economies in developing countries. *Procedia—Social and Behavioral Sciences, 15*, 2589–2594. Retrieved from http://www.sciencedirect.com/science/article/pii/S1877042811006975.

Zanini, N., & Benton, T. (2015). The roles of teaching styles and curriculum in mathematics achievement: Analysis of TIMSS 2011. *Research Matters, 20*, 35–44.

Chapter 3
Methodology: Examining
the Globalization of Science Curricula
Using TIMSS

Abstract Twenty years of collected TIMSS data reveals interesting insights into the globalization of science curricula. To answer the research questions, three different methods were used to analyze the TIMSS dataset. First, changes in countries' intended science curricula were captured and coded over the course of three TIMSS cycles (1999, 2007 and 2015 for Grade 8, and 2003, 2007 and 2015 for Grade 4). Changes were identified using countries' responses to the TIMSS curriculum questionnaires. This approach tracks changes in national science curricula over time. Second, cluster and discriminant analyses of the curriculum questionnaire data were used to determine potential convergence of curricula; countries may be clustered into groups on the basis of the topics included or not included in their intended science curricula. Third, the TIMSS encyclopedias and TIMSS teacher questionnaires provide detailed information on additional features of the implemented science curricula, such as the mean time spent teaching science in each country, or the percentage of students taught the TIMSS science topics. Such information was carefully analyzed for a subsample of countries. However, inconsistencies in the way this information was collected and presented across different TIMSS cycles made comparisons across countries between cycles challenging. As each of the three methods has its advantages and limitations, investigating the three research questions from different angles and with different, yet complementary techniques offers the most comprehensive analysis of the available TIMSS data.

Keywords Cluster analysis · Curriculum coding · Curriculum convergence Discriminant analysis · International large-scale assessment · Science education Trends in Mathematics and Science Study (TIMSS) · TIMSS curriculum questionnaire

A review of the existing literature identified several methods have been used previously to investigate globalization of curricula The research methodology we used to investigate our three research questions can be separated into three distinct strands.

© International Association for the Evaluation of Educational Achievement (IEA) 2018   23
O. Stacey et al., *The Globalization of Science Curricula*, IEA Research
for Education 3, https://doi.org/10.1007/978-3-319-71532-2_3

(1) We coded responses to the TIMSS science curriculum questionnaire data, in order to track changes in national science curricula over time.
(2) We undertook cluster and discriminant analyses, which grouped countries into distinct groups on the basis of the TIMSS science topics included in their intended science curricula. (The reasons and rationale for using cluster and discriminant analyses over other statistical techniques such as latent class analysis will be discussed later in the chapter.)
(3) We coded additional science curriculum features for a smaller sample of countries.

All three methods made use of data collected by the TIMSS curriculum questionnaire, which provides information on which TIMSS science topics are included in the intended science curricula of participating countries. In each case, we elaborate further on the limitations of the data sources and methods.

3.1 Coding of Curriculum Questionnaire Data

The TIMSS curriculum questionnaires provide a rich array of data on the science curricula and science teaching in participating countries at Grades 4 (ages 9–10) and 8 (ages 13–14). One section of the curriculum questionnaires pertains to science topics covered at each particular grade (either Grade 4 or 8). The questionnaire asks respondents (typically representatives from the Ministry of Education or equivalent) to indicate whether each TIMSS science topic is taught to "all or almost all students" in the grade, "only the more able students", or whether it is "not included in the curriculum" through to that grade (see Fig. 3.1 for an example from the questionnaire).

The responses to the questionnaires reveal the TIMSS science topics included in a country's science curriculum for a particular grade in a particular year. To address our first research question on whether there have been changes in countries' intended science curricula since 1999, individual countries' responses to the curriculum questionnaires between 1999 and 2015 were coded.

By comparing individual countries' responses to the curriculum questionnaire between different TIMSS cycles, changes in the science topics included in countries' intended science curricula over time can be detected.

Because the TIMSS curriculum questionnaire was first administered in 1999 for Grade 8 and 2003 for Grade 4, it was not feasible to code responses from every cycle of TIMSS; TIMSS 1999 was used as the baseline for Grade 8 and TIMSS 2003 as the baseline for Grade 4 (Table 3.1). For both grades, TIMSS 2015 was the most recent time point that could be considered. In addition to the earliest and most recent TIMSS cycles with available curriculum questionnaires, an intermediate TIMSS cycle (2007) was also included in this coding exercise. The inclusion of an intermediate cycle allowed changes in a country's science curriculum to be detected at more than one time point. This is important as countries may have changed the

Eighth Grade Science Topics Covered

This science module refers to the national curriculum that was in effect for the eighth grade students assessed in TIMSS 2015—the curriculum that covers science instruction at the eighth grade of formal schooling for the majority of students. If you do not have a national curriculum, please summarize for your state or provincial curricula.

S8. (i) According to the national science curriculum, what proportion of grade 8 students should have been taught each of the following topics or skills by the end of grade 8?

Be sure to include curriculum expectations for all grades up to and including grade 8. Grades represent years of formal schooling. For example, if "Year 9" in your country corresponds to the eighth year of formal schooling, please choose grade 8.

(ii) Across grades from preprimary through upper secondary education, at what grade(s) are the topics primarily intended to be taught?

If there are not any specifications to this detail, please indicate national expectations to the best of your ability. If part of a topic does not apply [e.g., energy flow in part A topic (f)], please explain in the comment field.

A. Biology	(i) Proportion of grade 8 students expected to be taught topic			(ii) Grade(s) topic is expected to be taught preprimary (PP) through the end of upper secondary (G12)												
	*Check **one** circle for each line.*			*Check the corresponding grade(s) for each topic*												
	All or almost all students	Only the more able students	Not included in the curriculum through grade 8	PP	G1	G2	G3	G4	G5	G6	G7	G8	G9	G10	G11	G12
a) Differences among major taxonomic groups of organisms (plants, animals, fungi, mammals, birds, reptiles, fish, amphibians)	○	○	○	▢	▢	▢	▢	▢	▢	▢	▢	▢	▢	▢	▢	▢
b) Major organs and organ systems in humans and other organisms (structure/function, life processes that maintain stable bodily conditions)	○	○	○	▢	▢	▢	▢	▢	▢	▢	▢	▢	▢	▢	▢	▢
c) Cells, their structure and functions, including respiration and photosynthesis as cellular processes	○	○	○	▢	▢	▢	▢	▢	▢	▢	▢	▢	▢	▢	▢	▢
d) Life cycles, sexual reproduction, and heredity (passing on of traits, inherited versus acquired/learned characteristics)	○	○	○	▢	▢	▢	▢	▢	▢	▢	▢	▢	▢	▢	▢	▢
e) Role of variation and adaptation in survival/extinction of species in a changing environment (including fossil evidence for changes in life on Earth over time)	○	○	○	▢	▢	▢	▢	▢	▢	▢	▢	▢	▢	▢	▢	▢
f) Interdependence of populations of organisms in an ecosystem (e.g., energy flow, food webs, competition, predation) and factors affecting population size in an ecosystem	○	○	○	▢	▢	▢	▢	▢	▢	▢	▢	▢	▢	▢	▢	▢
g) Human health (causes of infectious diseases, methods of infection, prevention, immunity) and the importance of diet and exercise in maintaining health	○	○	○	▢	▢	▢	▢	▢	▢	▢	▢	▢	▢	▢	▢	▢

Fig. 3.1 Extract from curriculum questionnaire for Grade 8 science TIMSS 2015. *Source* TIMSS 2015 Grade 8 curriculum questionnaire © 2015 International Association for the Evaluation of Educational Achievement (IEA). Publisher: TIMSS & PIRLS International Study Center, Lynch School of Education, Boston College

Table 3.1 TIMSS cycles used in the analysis

Grade	TIMSS cycles used
4	2003, 2007, 2015
8	1999, 2007, 2015

topics in their science curriculum more than once between the baseline and 2015 cycles. If a science topic was added between 1999 and 2007 and then removed between 2007 and 2015, the change would not be detected if only the baseline and 2015 cycles were considered.

The science topics included in the curriculum questionnaire varied slightly for each cycle of TIMSS. However, in order to map changes in the participating countries' curricula consistently across the TIMSS cycles, it was necessary to consider the same set of science topics in each cycle. Therefore, we first mapped the science topics across the TIMSS cycles to identify equivalent topics in the curriculum questionnaires of each TIMSS cycle.

We used the TIMSS 2015 science topics as the basis for this mapping exercise, matching science topics in the older TIMSS curriculum questionnaires to the 2015 TIMSS science topics. In some cases, there was a direct match in science topics across cycles, while, in other cases, there was no match. We found there were science topics in the TIMSS 2015 curriculum questionnaire that were not included in earlier questionnaires and, similarly, there were topics from earlier cycles that were not included in the 2015 curriculum questionnaire. As an illustration, the 1999 questionnaires included topics on the nature of science and scientific inquiry, but there were no equivalent science topics in the 2015 questionnaire. Only science topics that could be matched across all three TIMSS cycles were included in the coding exercise. For Grade 8 there were 20 science topics that mapped across the three TIMSS cycles and for Grade 4 there were 21 such topics (for a full list of these topics and the results of this mapping, see Chap. 4, Sect. 4.1.5).

To identify the nature of the curriculum change, we developed a coding framework (Table 3.2).

Table 3.2 Codes used in the curriculum mapping exercise

Code	Description of code
−2	Science topic has been removed from the curriculum at that grade
−1	Reduced emphasis on the science topic in the curriculum at that grade (topic was previously taught to most or all students in grade, but is now only taught to the more able students in the grade)
0	No change to science topic
1	Greater emphasis on the science topic in the curriculum at that grade (topic was previously taught only to more able at that grade but is now taught to most or all students in the grade)
2	Science topic has been introduced to the curriculum at that grade

In order to complete the coding exercise, the responses of individual countries to the curriculum questionnaire for each TIMSS cycle considered were obtained from the IEA data repository (data and documentation files from completed IEA studies are freely available for research purposes at http://www.iea.nl/data). For each science topic that mapped across the three TIMSS cycles, a country's response to whether the science topic was intended to be taught to "all or most students", "only the more able students", or was "not included in the curriculum" through to that grade was compared between cycles. For example, in Iran in 2003 at Grade 4, the curriculum questionnaire response to the life sciences topic "the characteristics of living things and the major groups of living things" indicated that the topic was taught to "all or most students". However, in 2007, the curriculum questionnaire response for this topic indicated that it was "not included in the curriculum through Grade 4". Therefore between 2003 and 2007 this topic was coded as a -2 (science topic removed from the curriculum at that grade) for Iran. This process was repeated for each country that had participated in at least two of the TIMSS cycles under consideration. Pairwise comparisons were made, for example at Grade 4, comparing the curriculum questionnaire responses in 2003–2007, 2007–2015, and 2003–2015.

This method addresses our first research question concerning whether there have been changes in countries' intended science curricula, as well as providing information about the nature of these changes; for example, whether countries were increasing the breadth of their curriculum by adding additional topics. This approach also partially addresses our second research question about whether science curricula are becoming increasingly similar across countries over time, as the coding indicates whether these changes are leading to an increased convergence in the intended science curricula across countries over time. The analysis also provides some evidence for our third research question on the potential existence of an international core science curriculum, by identifying science topics that are included in the intended curricula of all or the majority of countries participating in TIMSS.

3.2 Cluster Analysis and Discriminant Analysis

Our second research question asked whether science curricula taught in different countries have been becoming more similar over the last 20 years. We investigated an array of complementary statistical techniques, and opted for a cluster analysis followed by a discriminant analysis.

We applied the cluster analysis to the first TIMSS cycle (1999 for Grade 8 and 2003 for Grade 4) in order to identify clusters of countries that were similar in terms of their intended science curricula. The cluster analysis was conducted on countries' responses to the section of the curriculum questionnaire detailing which science topics were included in the intended curriculum at each grade. Only the science topics which could be mapped across all three TIMSS cycles were included in the analysis to ensure comparability. The second step, the discriminant analysis, was again performed on the first cycle of TIMSS, in order to produce a model that was then used to classify countries in the following TIMSS cycles.

The advantage of this two-step technique over a simple cluster analysis repeated on the three cycles of TIMSS considered is that, while the cluster analysis classifies countries based on the specific information of what is taught in each of the three cycles using the exact same model every time, the discriminant analysis uses a completely separate algorithm to classify countries in each of the three cycles.

In this investigation, the comparison of clusters of countries over time is not meaningful as the clusters will have been created using different criteria each time. Furthermore, the set of countries participating in each cycle is not constant, with new countries joining and some countries which participated in the early TIMSS cycles leaving. This issue is partially addressed by discriminant analysis. This is because new entrants to the more recent TIMSS cycles do not have an effect on the clustering rule, which is fixed. As a result, new entrants are simply classified as belonging to one cluster group based on their responses to the TIMSS curriculum questionnaire.

That said, the clustering algorithm itself is influenced by the specific set of countries that were included in the first TIMSS cycle. A straightforward solution to this problem would be to restrict the analysis to countries that have participated in all three TIMSS cycles. We chose not to pursue this approach as we considered the advantages to be more than offset by the disadvantages.

One of the main drawbacks to this approach is that it would narrow the analysis to a relatively small set of countries (21 for Grade 8 and 18 for Grade 4) and, as a consequence, would limit the conclusions we would be able to make from the analysis. This, in turn, would give a very partial response to our research question, as we would be testing the hypothesis of convergence in science curricula on a subsample of countries. What is going on in this specific subsample might be not representative of what is happening in the overall population. With this caveat in mind, we chose to prioritize a more complete analysis and exploit all the information available by running our model on the full sample.

A limitation of our approach is that the number of clusters is identified using the first TIMSS cycle and is therefore fixed for the following two cycles. If science curricula are becoming more dissimilar, this would result in a higher number of clusters in more recent TIMSS cycles. We would not be able to detect this phenomenon using our approach. However, given the relative stability of science curricula, this event is unlikely. To check this assumption, when we conducted three separate cluster analyses on the three cycles of TIMSS, the ideal number of clusters identified by the algorithm was always the same (two in each case).[1]

---

[1]The results of the three separate cluster analyses are not reported due to concerns regarding the misinterpretation of the outcome. Nevertheless, the finding that the number of groups identified by the algorithm is constant in all cycles is relevant in establishing that the science curricula of countries are not diverging to such an extent to justify the identification of an increasing number of clusters over time.

Having outlined our statistical approach and the rationale behind it we now give a more thorough account of the statistical techniques used in this investigation.

Cluster analysis is a statistical technique that allows the grouping of a set of observations according to certain characteristics, in such a way that observations in the same cluster are more similar to each other than to observations in other groups. In the context of our investigation, the observations that are clustered are countries, and the variables used to group them are the responses given to the TIMSS curriculum questionnaire items on the science topics that are included in countries' intended science curricula. Our analysis aims to investigate the existence of convergence in science curricula, with convergence signaled by the tendency of a group to expand at the expense of others.

When running the analysis on the TIMSS dataset we chose to implement a two-step cluster analysis. The first step of this procedure consists of pre-clustering the records into small sub-clusters that are then consolidated into a smaller number of groups in the second step. The advantage of this methodology over other clustering algorithms is that, when the number of groups is unknown, the algorithm automatically returns the optimal number of clusters based on the Bayesian information criterion (BIC).[2] Observations are grouped together based on a distance measure that, in this case, reflects how different countries are from each other in terms of science topics covered in their curricula. In order to cluster the observations, we adopt the log-likelihood criterion that is appropriate when grouping observations using continuous as well as categorical variables. This distance measure works best when all variables are independent and categorical variables have a multinomial distribution.[3] The log-likelihood is a probability-based distance. The distance between two clusters is related to the decrease in log-likelihood as they are combined.[4]

Our original intention was to perform a latent class analysis on the data. We performed a cluster analysis followed by a discriminant analysis due to the reduced size of the sample. Given the large number of parameters estimated, latent class analysis requires a considerable number of observations. (In statistics, a model cannot be identified unless the sample size is one more than the number of predictors and, as a rule of thumb, we need around ten observations per parameter to estimate a model with reasonable precision.)

---

[2]The Bayesian information criterion (BIC) is used to select between different models based on how well they fit the data and the number of parameters they use. The BIC is an increasing function of both the number of parameters (the BIC penalizes the complexity of the model) and the error variance (a measure of how well the model fits the data). The algorithm chooses the model with the lowest BIC that implies either fewer explanatory variables, better fit, or both.

[3]In practice, these conditions are rarely satisfied, but the algorithm is thought to behave reasonably well when the assumptions are not met.

[4]The goal of the clustering algorithm is to maximize the probability (or likelihood) of obtaining the figures observed in the dataset, given the clusters that are identified at the final stage. Hence, given that our aim is to maximize the likelihood function, the bigger the drop that results from merging two clusters, the more dissimilar the observations therein contained.

Previous studies which have used TIMSS data to categorize countries' curricula into particular groups using latent class analysis, such as Zanini and Benton (2015), made use of teacher questionnaire responses as opposed to curriculum questionnaire responses. There are a far greater number of observations in the teacher questionnaire than there are in the curriculum questionnaire, making the latent class analysis approach viable for studies using teacher questionnaire data. It is not possible to use latent class analysis in this study as there is only one observation available for each participating country, and this study uses the curriculum questionnaire as opposed to the teacher questionnaire because of time and resource constraints. That said, the purpose of cluster analysis is the same as latent class analysis but, instead of using a model to group observations, it employs distance measures that are less demanding in terms of number of observations needed.

Discriminant analysis is, in a sense, used as a tool to reverse what has been implemented with cluster analysis. It is a statistical technique used to determine which variables discriminate between two or more groups (Ayinla and Adekunle 2015). This procedure was performed on the groups of countries that had been previously identified by the cluster analysis. It created a model which identified the most important science topics in determining the grouping of countries.

To illustrate how this works in practice, if a particular science topic, for example "forces", is taught in all countries in group A and in none of the countries in group B, given group A and B,[5] the algorithm that performs discriminant analysis is able to detect that this science topic is an important element of discrimination between the two groups. The algorithm infers that countries teaching that specific science topic will be highly likely to belong to group A, while countries that do not teach the topic will be highly likely to belong to group B.

In reality, the algorithm performing discriminant analysis is slightly more complex, as it takes into account all the science topics at once and, for each of them, estimates a coefficient that reflects the probability of belonging to a given group based on the answer given to that particular question (for example, whether students at that grade are expected to be taught that science topic or not). When all the coefficients are fitted in a linear model, for each observation (country), we are able to predict the group identity based on the answers given in the curriculum questionnaire.

The model can then be used to predict the clustering of countries in subsequent years. If convergence of science curricula has occurred in the last 20 years, we should observe a tendency for countries to concentrate in a single group.

As indicated above, this approach has advantages compared to cluster analysis alone but it is not immune from criticism. Firstly, the approach enables the identification of convergence (when one group of countries grows over time at the expense of others). However, as already mentioned, it does not enable the detection of divergence. This shortfall arises because, by construction, the number of clusters in cycles following the baseline year will be equal to or lower than the number of

---

[5]In this case the two groups were identified by means of cluster analysis.

clusters detected in that baseline year. We have already provided explanations for why this is not a major concern in our specific dataset. Moreover, we should point out that the scope of this study is to assess whether convergence in science curricula has occurred rather than divergence.

A second concern is that, although this procedure allows the detection of whether science curricula are getting more similar over time, it also focuses on features that made the science curricula of countries different in the first place. This means that the science topics that are particularly important in determining the group a country belongs to in the baseline year will keep on playing a primary role in subsequent cycles. Additionally, science topics that do not discriminate strongly between groups in the baseline cycle will only play a marginal role in the classification of countries in subsequent years.

Continuing with the earlier example, if, in the baseline year, all countries in group A are teaching a given science topic and no countries in group B are teaching that topic, but, in the second cycle, all countries include the science topic in their curriculum, the algorithm would, in this case, predict that all observations in the second cycle belong to group A. In light of this result, we could be tempted to conclude that curricula are converging but this might not be the case. If there is another science topic in the questionnaire and this topic is taught by all countries in the first cycle, the algorithm would completely ignore this item as it is not able to discriminate between observations (countries). However, if in the following cycle, some countries stop teaching that topic so that part of the sample has it in their science curricula and part of the sample does not, the model will not be able to exploit this information because the topic is not included in the predictive model.[6] In this example, one group grows at the expense of the other, but convergence is not supported. The curricula are becoming more similar in some dimensions, but are diverging in other aspects that our model is not taking into account.

This is only a hypothetical example and, in reality, the situation is more complicated as the algorithm considers multiple science topics at a time. Moreover, it is unlikely to come across science topics with zero predictive power.[7] More often we will have science topics with low predictive power (i.e. not able to discriminate between groups with high precision). In this case, the information will not be completely disregarded but the algorithm will instead attach a low coefficient to the science topic. As a consequence the answer given to the question related to that science topic will affect the probability of belonging to a certain group only by a very small degree.

In order to resolve the issue related to the dependence of our results on the predictive power of science topics in the initial cycle of the questionnaire, the same process was repeated in reverse. This means that, starting from the most recent TIMSS

---

[6]We could consider this topic to be included in the model with a zero coefficient, reflecting the fact that its predictive power was null in the first cycle.

[7]This occurs if, as in our example, all countries in the sample give the same answer to a certain question.

cycle available (2015 for both Grade 4 and Grade 8), the data were clustered and a discriminant analysis performed based on the resulting groups. Using the model that emerges from the discriminant analysis, countries were classified according to the answers given in earlier cycles of the questionnaire. This reverse approach was used to assess differences in curricula based on science topics that were homogeneously taught (or not taught) in the first TIMSS cycle and hence have diverged.

Using the previous example, with this new approach, the algorithm will now cluster countries based on the answers given in the final cycle of the questionnaire. A science topic that did not discriminate in the first cycle (as it was taught by all countries) is now an important predictor for clustering observations in the final cycle, and the science topic that was a good discriminant in first place, is no longer relevant. Applying the reverse approach to this case, the cluster analysis conducted on the final cycle will group observations into two separate groups while assigning all observations to a single group in the initial cycle (because all countries gave the same answer to the item in the earlier cycle). Given that one of the clusters was bigger in the first cycle than in the end-point questionnaire, we can conclude that curricula have diverged along the specific dimension (science topic) we are analyzing. The interpretation will be complicated by the fact that in reality our model considers multiple science topics at a time.

It is important to note that discriminant analysis is a linear technique and we have applied it to a dataset in which we have categorical variables. This approach is not inappropriate provided the underlying assumptions are understood. The challenge of applying a linear technique to categorical variables is that, for each item, the model will treat the categories as continuous variables and will return a single coefficient that has the same interpretation as the beta in a linear regression.[8] Given that our variables are measured on a three-point scale (ranging from 1 to 3), for a given science topic, the probability of belonging to a given group will be two times larger/smaller for a country with a value of 2 (topic taught to only the more able students) compared to a country with a value of 1 (topic taught to all or most students). Similarly, given the linear nature of the coefficients, the probability of belonging to a given group will be three times larger/smaller for a country with a value of 3 (topic not taught in a given grade) compared to a country with a value of 1 (topic taught to all or most students). This is a limitation of our analysis but, as long as we are willing to assume that individuals in the middle of the scale (i.e. teaching to the more able students only) are equally different from individuals at the two extremes (teaching to all or teaching to nobody), then the use of a linear model is appropriate. For our purpose the assumption is not unreasonable considering the fact that the limited amount of data prevents us from estimating a non-linear model.[9]

---

[8]In a linear regression the beta coefficient represents the difference in the predicted value of Y for each one-unit difference in a certain variable, keeping all other variables constant.

[9]Here we encounter a similar problem to the one we have already discussed for latent class analysis. A non-linear model (such as latent class analysis) would require the estimation of a coefficient for each of the values taken by each categorical variable. In our case, for example at Grade 8, we have 20 variables with three possible outcomes each, implying 40 coeffients to estimate. This is impossible given the sample size.

From this short description of our statistical analysis it is evident that assessing convergence in science curricula is not a straightforward task and that different techniques are likely to result in different outcomes. Given that each approach has advantages and disadvantages, investigating the research question from different angles and with different yet complementary techniques offers the most robust solution.

Finally, readers should be aware that cluster analysis does not involve hypothesis testing and the computation of significance levels. As a result, the appropriateness of a solution can only be assessed by the researcher. This should be evaluated through the critical observation of the elements composing the clusters and the difference in the pattern of answers given by countries belonging to different groups.

3.3   Additional Analysis of a Sample of Countries

Both the coding of curriculum questionnaires and statistical methods outlined in Sects. 3.1 and 3.2 make exclusive use of responses from the TIMSS curriculum questionnaire. This allows a rigorous comparison of the intended science curricula of countries participating in TIMSS over time. This dataset, however, gives little information about the implemented science curriculum of a country. To complement these two earlier approaches, a number of other features of the implemented science curriculum were captured for a selection of countries that have participated in at least two of the TIMSS cycles considered in this report. Changes in these features were recorded to investigate whether there is convergence over time.

We collected information about various aspects of the implemented science curricula, such as the mean time spent teaching science in each country, the percentage of students taught the TIMSS science topics, the organization and structuring of the science curriculum, and science assessment. This information came from the TIMSS encyclopedias and from the TIMSS teacher questionnaires for the selected countries (Martin et al. 2000, 2004, 2008; Mullis et al. 2016). For this part of our study, we selected 15 countries at Grade 4 (Table 3.3) and 16 countries at Grade 8 (Table 3.4) on the basis of a number of different criteria. Firstly, the majority of these countries have appeared in all three TIMSS cycles under study at both Grade 4 and 8. The countries that have not, such as Qatar, have appeared in at least two cycles and were included to widen the geographic distribution and achievement profile of the sample. In a few cases, a country that participated in all three TIMSS cycles was not included. This was either because the country appeared in three cycles for one grade but not the other, or because there was no data available from the curriculum questionnaire on the topics included in the intended science curriculum in a particular grade. This was the case, for example, for the Netherlands. Such countries were discounted from the selection. Countries which participated in only one relevant TIMSS cycle were also excluded from this analysis, thus ensuring that the amount of information that can be collected on each country was maximized.

Table 3.3 Countries included in additional analysis at Grade 4

Country	Average TIMSS science score in 2003	Average TIMSS science score in 2007	Average TIMSS science score in 2015
Morocco	304 (6.7)	297 (5.9)	352 (4.7)
Iran	414 (4.1)	436 (4.3)	421 (4.0)
Qatar	N/A	294 (2.6)	436 (4.1)
New Zealand	520 (2.5)	504 (2.6)	506 (2.7)
Italy	516 (3.8)	535 (3.2)	516 (2.6)
Australia	521 (4.2)	527 (3.3)	524 (2.9)
Lithuania	512 (2.6)	514 (2.4)	528 (2.5)
England	540 (3.6)	542 (2.9)	536 (2.4)
Hungary	530 (3.0)	536 (3.3)	542 (3.3)
Slovenia	490 (2.5)	518 (1.9)	543 (2.4)
United States	536 (2.5)	539 (2.7)	546 (2.2)
Chinese Taipei	551 (1.7)	557 (2.0)	555 (1.8)
Hong Kong	542 (3.1)	554 (3.5)	557 (2.9)
Japan	543 (1.5)	548 (2.1)	569 (1.8)
Singapore	565 (5.5)	587 (4.1)	590 (3.7)

Note Standard errors are provided in brackets after the average achievement scores. *N/A* not applicable

   In our selection of countries, we endeavored to reflect the geographic scope of TIMSS and also represent the full range of achievement on the TIMSS assessment. This means that the sample includes high achievers such as Singapore and Japan, more moderate achievers such as Italy and New Zealand, and countries that have performed less well, such as Morocco, Qatar and Iran. Furthermore, our sample includes Middle Eastern countries, East Asian countries, European countries, southern hemisphere nations and African nations. The selection also includes some less developed countries, such as Iran, and takes into account the outcomes of the cluster analysis by including countries from each cluster identified by the analysis.

   The same 15 countries were selected for both Grade 4 and Grade 8 as this represented the most efficient strategy for collecting information on additional aspects of science curricula, such as time allocated to science and percentage of students taught the TIMSS science topics. For Grade 8, Israel was included as a sixteenth country as the literature review identified it as a country that had made extensive changes to its science curriculum as a result of participation in TIMSS (Klieger 2015). As a consequence, exploring the science curriculum in Israel seems highly relevant to this study.

Table 3.4 Countries included in additional analysis at Grade 8

Country	Average TIMSS science score in 1999	Average TIMSS science score in 2007	Average TIMSS science score in 2015
Morocco	323 (4.3)	402 (2.9)	393 (2.5)
Iran	448 (3.8)	459 (3.6)	456 (4.0)
Qatar	N/A	319 (1.7)	457 (3.0)
Italy	493 (3.9)	495 (2.8)	499 (2.4)
Israel	468 (4.9)	468 (4.3)	507 (3.9)
Australia	540 (4.4)	515 (3.6)	512 (2.7)
New Zealand	510 (4.9)	N/A	513 (3.1)
Lithuania	488 (4.1)	519 (2.5)	519 (2.8)
Hungary	552 (3.7)	539 (2.9)	527 (3.4)
United States	515 (4.6)	520 (2.9)	530 (2.8)
England	538 (4.8)	542 (4.5)	537 (3.8)
Hong Kong	530 (3.7)	530 (4.9)	546 (3.9)
Slovenia	533 (3.2)	538 (2.2)	551 (2.4)
Chinese Taipei	569 (4.4)	561 (3.7)	569 (2.1)
Japan	550 (2.2)	554 (1.9)	571 (1.8)
Singapore	568 (8.0)	567 (4.4)	597 (3.2)

Note Standard errors are provided in brackets after the average achievement scores. *N/A* not applicable

3.4   Methodological Limitations

There are a number of inherent limitations in our methodological approach and these should be taken into account when interpreting the results.

Firstly, the primary data source used for coding changes in the intended science curriculum, and for the cluster analysis, is responses to the section of the curriculum questionnaire on science topics intended to be taught. We have assumed that these questionnaires have been completed accurately but have no way of testing our assumption. There are, however, likely to be sources of error in the completion of the questionnaires. For example, the science topics in each country's curriculum are unlikely to correspond word for word to the science topics listed in the TIMSS curriculum questionnaire. Therefore, it is likely that there is some degree of subjectivity in how the curriculum questionnaires are interpreted and completed in each country.

Secondly, we have assumed that the curriculum questionnaire responses for each country accurately reflect the intended curriculum for the whole country. Whilst this is likely to be the case for countries with highly prescribed statutory national science curricula, such as England, this may not be the case in all countries, such as those with a federal structure like Germany, or those with a decentralized curriculum.

A third limitation is that coding of changes in the curriculum approach can only be used in countries which have taken part in at least two of the TIMSS cycles considered. This excludes countries that only started participating in TIMSS in more recent cycles and also excludes the many developing countries that do not participate at all. This means that developed countries are over-represented in this analysis, which limits the generalizability of the conclusions.

Fourthly, the TIMSS science topics included in the curriculum questionnaires in each cycle change slightly. As noted earlier, in the TIMSS 1999 curriculum questionnaire there was a section on science topics based on the nature of science and scientific inquiry skills. There is no equivalent section in the 2015 curriculum questionnaire. It is consequently not possible to code for changes to these areas of the curriculum and it is likely that, as a result, our coding will not capture all the changes in the content and intended science curricula of countries over time.

Another limitation is that some of the science topics included in the curriculum questionnaire are likely to be covered in subjects other than science in some countries. For example, in England, some of the Earth science topics are taught in geography. Another limitation is that the time frame considered differs for Grade 4 and Grade 8. Ideally the same time frame would have been used for both grades, but this was not possible as TIMSS was not administered at Grade 4 in 1999. As a result, and to maximize the time frames explored in the study, different starting points were used for Grades 4 and 8.

There is the further challenge that in each TIMSS cycle there are different numbers of participating countries, with some countries dropping out and new countries entering. This makes direct comparisons between different cycles less straightforward and the interpretation of results more complex. Furthermore, the scope of this study does not allow us to explore the reasons why countries leave the TIMSS study or choose to enter it.

In addition, by using TIMSS curriculum questionnaire data, this study is restricted to considering globalization in science curricula from 1999 to 2015 for Grade 8 and 2003–2015 for Grade 4. This means that changes in science curricula prior to 1999 that may have led to increased global alignment in science curricula across countries will be missed. Other studies investigating globalization of curricula have encountered similar issues. Rutkowski and Rutkowski (2009) suggested that using data from the 1990s as a baseline to measure curricular change and global alignment may be too late to detect globalization in national curricula. They argued that globalization was already exerting an influence on curricula well before this point in time.

Despite these limitations we consider the curriculum questionnaire to be a very useful and rich source of information on the intended curriculum in countries participating in TIMSS. These datasets represent the largest comparative body of evidence on science curricula recorded across both a wide range of countries and over an extended time period. By combining analyses with in-depth examination of the features of the science curriculum for a subset of countries from other TIMSS sources, we also consider some aspects of countries' implemented science curricula, uncovering additional evidence for or against convergence over time.

References

Alayande, S., & Adekunle, B. (2015). An overview and application of discriminant analysis in data analysis. *IOSR Journal of Mathematics, 11*(1), 12–15. Retrieved from http://www.iosrjournals.org/iosr-jm/papers/Vol11-issue1/Version-5/B011151215.pdf.

Klieger, A. (2015). Between two science curricula: The influence of international surveys on the Israeli science curriculum. *The Curriculum Journal, 26*(3), 404–424.

Martin, M. O., Mullis, I. V. S., Gonzalez, E. J., Gregory, K. D., Smith, T. A., Chrostowski, S. J., et al. (2000). *TIMSS 1999 international science report, findings from IEA's repeat of the third international mathematics and science study at the eighth grade.* Chestnut Hill, Massachusetts: TIMSS & PIRLS International Study Center, Boston College. Retrieved from https://timssandpirls.bc.edu/timss1999i/pdf/T99i_Sci_All.pdf.

Martin, M. O., Mullis, I. V. S., Gonzalez, E. J., & Chrostowski, S. J. (2004). *Findings from IEA's trends in international mathematics and science study at the fourth and eighth grades.* Chestnut Hill, Massachusetts: TIMSS & PIRLS International Study Center, Boston College. Retrieved from https://timss.bc.edu/timss2003i/scienceD.html.

Martin, M. O., Mullis, I. V. S., Foy, P., Olson, J. F., Erberber, E., Preuschoff, C., et al. (2008). *TIMSS 2007 international science report.* Chestnut Hill, Massachusetts: TIMSS & PIRLS International Study Center, Boston College. Retrieved from https://timss.bc.edu/TIMSS2007/PDF/TIMSS2007_InternationalScienceReport.pdf.

Mullis, I. V. S., Martin, M. O., Foy, P., & Hooper, M. (2016). *TIMSS 2015 international results in science. Chestnut Hill,* Massachusetts: TIMSS & PIRLS International Study Center, Boston College. Retrieved from http://timssandpirls.bc.edu/timss2015/international-results/.

Rutkowski, L., & Rutkowski, D. (2009). Trends in TIMSS responses over time: Evidence of global forces in education? *Educational Research and Evaluation, 15*(2), 137–152.

Zanini, N., & Benton, T. (2015). The roles of teaching styles and curriculum in mathematics achievement: Analysis of TIMSS 2011. *Research Matters, 20,* 35–44.

Chapter 4
Results: Evidence for the Globalization of Science Curricula from TIMSS

Abstract In TIMSS, participating education systems are asked in the curriculum questionnaire which TIMSS science topics are included in their intended science curricula. By coding the changes in countries' responses, some measure of the changes in national science curricula between 1999 and 2015 (for Grade 8) and between 2003 and 2015 (for Grade 4) may be determined. Countries can also be categorized by how stable or changeable their curriculum is on the basis of the number of changes they have made to their curriculum, the average number of curriculum changes at Grades 4 and 8 for each TIMSS cycle and the topics common to the vast majority of participating countries' intended science curricula. The analysis found there was considerable variation among countries in terms of the number and nature of changes they made to their science curriculum between 1999 (2003 for Grade 4) and 2015. Cluster analysis and discriminant analysis revealed which science topics were most discriminating in terms of predicting country clusters. For both grades, and for each TIMSS cycle, the analyses produced two groups of countries based on responses to the curriculum questionnaire. The cluster and discriminant analyses provide stronger evidence for convergence in science curricula over time at Grade 8 than at Grade 4. The implemented science curriculum of 15 countries was examined in more depth using the TIMSS teacher questionnaire responses; due to difficulties in obtaining robust comparable data for countries across the different TIMSS cycles and the significant caveats that would have applied to any conclusions drawn from these analyses, these data are not considered.

Keywords Cluster analysis · Curriculum change · Curriculum convergence Discriminant analysis · International large-scale assessment · Science curriculum Science education · Science topics · Trends in Mathematics and Science Study (TIMSS)

© International Association for the Evaluation of Educational Achievement (IEA) 2018   39
O. Stacey et al., *The Globalization of Science Curricula*, IEA Research
for Education 3, https://doi.org/10.1007/978-3-319-71532-2_4

4.1   Coding of Curriculum Questionnaire Data Results

4.1.1   The Extent of Change in Countries' Science Curricula Over Time

By coding the changes in countries' responses to which TIMSS science topics are included in their intended science curricula, it was possible to identify the extent to which countries' intended science curricula had changed between 1999 and 2015 (for Grade 8) and between 2003 and 2015 (for Grade 4). We found considerable variation in the extent to which countries' intended science curricula had changed. At both Grade 4 and Grade 8, a number of distinct groups of countries could be identified from the data. These ranged from highly stable (fewer than three changes to TIMSS science topics included in countries' intended science curricula over the period of comparison) through to highly changeable (more than ten changes to TIMSS science topics included in countries' intended science curricula over the period of comparison).

In this coding exercise, a change to a science topic could either be its removal from the intended curriculum, its addition to the intended curriculum, or a change in emphasis for the science topic in the intended curriculum. A change in emphasis could be an increase in emphasis; for example, previously it may have been taught only to the more able students but is now intended to be taught to all or most students. Alternatively, it could be a reduction in emphasis; for example, previously intended to be taught to all or most students but now only intended to be taught to the more able students.

We coded these changes for all the countries that participated in both the 2003 and 2015 TIMSS cycles at Grade 4 (Table 4.1) and for all the countries that participated in both the 1999 and 2015 TIMSS cycles at Grade 8 (Table 4.2). For Grade 4, there were 21 TIMSS science topics included in this coding exercise

Table 4.1 Country categorizations for Grade 4 between 2003 and 2015

Category	Number of changes to science topics in intended curriculum	Number of countries or provinces in category	Countries or provinces in category
Highly stable	1–3	1	Singapore
Moderately stable	4–6	8	England, Lithuania, Russia, Italy, Hungary, Ontario, Slovenia, USA
Changeable	7–10	7	Armenia, Chinese Taipei, Cyprus, Hong Kong, Japan, New Zealand, Norway
Highly changeable	More than 10	4	Australia, Morocco, Belgium (Flemish), Quebec

Table 4.2 Country categorizations for Grade 8 between 1999 and 2015

Category	Number of changes to science topics taught in intended curriculum	Number of countries or provinces in category	Countries or provinces in category
Highly stable	1–3	7	Hungary, Hong Kong, Jordan, Turkey, USA, Iran, Chinese Taipei
Moderately stable	4–6	13	England, Japan, Malaysia, Slovenia, Thailand, Australia, Canada, Italy, Lithuania, New Zealand, Chile, Republic of Korea, Russia
Changeable	7–10	2	Singapore, Morocco
Highly changeable	More than 10	2	South Africa, Israel

whilst for Grade 8 there were 20 TIMSS science topics. The same comparison was made between 2003–2007 and 2007–2015 for Grade 4 and between 1999–2007 and 2007–2015 for Grade 8; in each case they yielded similar results.

The Singaporean curriculum was noticeably more stable than that of other countries at Grade 4 (Table 4.1). There was only one change in the TIMSS topics intended to be taught in Singapore at Grade 4 between 2003 and 2015, and the next most stable country (Lithuania) made only four changes to its intended science curriculum in the same period. By contrast, in Belgium (the country with the most changeable curriculum), there were 15 changes in the TIMSS topics intended to be taught. Given Singapore's high performance on TIMSS across the years it is unsurprising that the number of changes to the intended science topics taught was minimal between 2003 and 2015, as there was likely less incentive to revise the curriculum. By contrast, Morocco is one of the lowest performing countries in TIMSS science at Grade 4. It is likely that the high number of changes to its intended science curriculum have been instigated to improve future science performance generally, not just on TIMSS.

Similar patterns emerge from the Grade 8 data in terms of distinct groups of countries demonstrating varying number of changes to their intended science curriculum (Table 4.2). However, the countries in each group do not necessarily match the Grade 4 groupings. For example, although Singapore shows remarkable stability in its science curriculum at Grade 4, changes in the intended science curriculum at Grade 8 have been far more extensive. It is important here to note that, for Grade 4, the comparison is being made between 2003 and 2015, whereas for Grade 8 the comparison is between 1999 and 2015 and so covers a longer time period. Additionally, the countries taking part in TIMSS at Grade 4 and Grade 8 are not exactly the same.

4.1.2   The Nature of Change in Countries' Science Curricula

This analysis enabled the nature of changes to the science curriculum of each country to be identified in addition to the number of changes made. Whilst countries could be grouped into categories based upon the extent to which their curricula had changed or remained stable, the profile of countries within each grouping could be quite different. The analysis identified some countries as "cutters" and some as "adders". Cutters are countries where curricular changes resulted in fewer TIMSS science topics appearing in the intended science curricula, whereas adders are countries where curricular changes tended to result in additional TIMSS science topics appearing within the intended science curricula. A third group of countries was identified, which we decribe as "balancers". These are countries where curricular changes resulted in roughly equal numbers of TIMSS topics being added and removed from the intended science curricula. We identified countries for each of these categories for Grade 4 (Table 4.3, comparison between 2003 and 2015 TIMSS data) and for Grade 8 (Table 4.4, comparison between 1999 and 2015 TIMSS data). Similar patterns emerged when comparing the data for 2003–2007 and 2007–2015 at Grade 4 and the data for 1999–2007 and 2007–2015 at Grade 8 so, in the interests of space, only the 2003 and 1999–2015 results are presented here.

Whilst the same categories exist for both Grade 4 and Grade 8, the proportion of countries in each category varies for the two grades (Tables 4.3 and 4.4).

Table 4.3 Country categorization for Grade 4 (2003–2015)

Category	Number of countries or provinces in category	Countries or provinces
Adders	8	England, Belgium (Flemish), Hungary, Japan, Chinese Taipei, Morocco, Armenia, Cyprus
Cutters	6	Italy, Lithuania, Singapore, Norway, Australia, Ontario
Balancers	6	Russia, Slovenia, Hong Kong, New Zealand, Quebec, USA

Table 4.4 Country categorization for Grade 8 (1999–2015)

Category	Number of countries or provinces in category	Countries or provinces
Adders	13	South Africa, Israel, Chile, England, Japan, Thailand, Italy, Republic of Korea, Hong Kong, Iran, Chinese Taipei, Jordan, Malaysia
Cutters	8	Singapore, Australia, New Zealand, Russia, Lithuania, Slovenia, Turkey, Hungary
Balancers	3	Morocco, Canada, USA

Additionally, countries can have different profiles for Grade 4 and Grade 8. For example, at Grade 4, Morocco is categorized as an adder, as between 2003 and 2015 it added four science topics to its curriculum and removed one. By contrast, at Grade 8, Morocco is classed as a balancer, as between 1999 and 2015 it added four science topics and removed four science topics from its curriculum.

4.1.3 Changes to Science Curricula Between TIMSS Cycles

The analysis enabled the average number of curricular changes between TIMSS cycles to be calculated (Tables 4.5 and 4.6). As the number of countries participating in each TIMSS cycle varies, the number of countries within each comparison differs (Tables 4.5 and 4.6).

Table 4.5 Average number of changes to science curriculum at Grade 4

Comparison	Number of countries in comparison	Average number of topics added	Average number of topics removed	Average number of topics with increased emphasis	Average number of topics with reduced emphasis	Average number of topics with no change
2003–2007	21	2.95 (2.66)	1.76 (2.09)	0.57 (1.37)	0.05 (0.21)	15.67 (2.92)
2007–2015	18	3.44 (2.77)	2.39 (2.21)	1.06 (2.12)	0.53 (1.29)	12.50 (4.76)
2003–2015	26	3.08 (3.19)	2.12 (2.21)	0.23 (0.50)	0.73 (1.53)	14.88 (3.15)

Note Standard deviations for each average are given in brackets

Table 4.6 Average number of changes to science curriculum at Grade 8

Comparison	Number of countries in comparison	Average number of topics added	Average number of topics removed	Average number of topics with increased emphasis	Average number of topics with reduced emphasis	Average number of topics no change
1999–2007	25	1.88 (2.01)	1.64 (1.98)	0.32 (0.61)	0.52 (1.58)	15.8 (4.05)
2007–2015	36	2.08 (2.11)	1.44 (1.80)	0.39 (1.03)	0.67 (1.29)	15.42 (3.07)
1999–2015	23	2.65 (2.74)	1.39 (1.55)	0.17 (0.48)	0.61 (1.31)	15.13 (3.07)

Note Standard deviations for each average are given in brackets

For each comparison, at both grades, the average number of science topics added is greater than the average number of topics removed (Tables 4.5 and 4.6). This suggests that, across countries as a whole, changes in a country's science curriculum are more likely to lead to an increase rather than a decrease in the number of TIMSS science topics included in the curriculum.

It is also more common for topics to be completely removed or added to the science curriculum than it is for there to be a change in emphasis to a science topic in the curriculum. Additionally, the results suggest that the science curriculum at Grade 4 is more subject to change than the science curriculum at Grade 8, as the average number of topics added or removed in each comparison is higher for Grade 4 than Grade 8. This finding is consistent with our earlier observation that, at Grade 8, a higher proportion of countries are categorized as having highly stable or moderately stable curricula than at Grade 4 (Tables 4.1 and 4.2).

4.1.4   TIMSS Science Topics Included in Countries' Curricula

This analysis enabled us to identify the extent to which TIMSS science topics were included in the intended curricula of participating countries, and whether there were particular TIMSS science topics that were core to the majority of participating countries' science curricula.

As the science topics included in each TIMSS cycle changed slightly each cycle, the science topics from the earlier TIMSS cycles were mapped to the 2015 TIMSS science topics to identify those that were common across all three cycles.

The number of TIMSS science topics included in the majority (80% or more) and minority (50% or fewer) of countries' intended curricula for Grade 4 and Grade 8 change over time (Tables 4.7 and 4.8).

For Grade 4, between 2003 and 2015, the number of TIMSS science topics taught in at least 80% of participating countries increased, whilst the number of topics taught to fewer than half of participating countries decreased (Table 4.7). This suggests that there is a growing number of core science topics at Grade 4 that are included in the majority of participating countries' intended science curricula. However, caution needs to be exercised in interpreting these results, as the number

Table 4.7   Science topics taught in the majority and minority of participating countries at Grade 4

TIMSS year	Number of participating countries	Number of science topics covered by 80% or more of countries	Number of science topics covered by 50% or fewer of countries
2003	29	2	6
2007	46	5	3
2015	53	7	2

Table 4.8 Science topics taught in the majority and minority of participating countries at Grade 8

TIMSS year	Number of participating countries	Number of science topics covered by 80% or more of countries	Number of science topics covered by 50% or fewer of countries
1999	39	12	0
2007	58	12	0
2015	45	14	0

and range of countries (in terms of their geographic distribution and level of economic development) participating in TIMSS at Grade 4 has increased between 2003 and 2015 and the TIMSS science framework itself changes slightly between cycles.

At Grade 8 (Table 4.8), the number of TIMSS topics included in at least 80% of participating countries' curricula was much higher than for Grade 4. There was only a modest increase of two topics between 1999 and 2015 in the number of TIMSS topics included in at least 80% of participating countries' curricula. Additionally, the number of TIMSS science topics covered by fewer than 50% of participating countries at Grade 8 was less than at Grade 4, and remained at zero in all three TIMSS cycles. The same caveats apply for Grade 8 as Grade 4 when interpreting these results and making comparisons over time.

4.1.5 Core and Non-core TIMSS Science Topics Included in Science Curricula

In addition to identifying the number of TIMSS science topics that were included in a country's science curriculum, the coding enabled the identification of core and minority (non-core) science topics. We identified the core TIMSS science topics as those that were included in the intended science curricula of 80% or more of participating countries at Grade 4 in 2003, 2007 and 2015, as well as the topics that appeared less frequently in the intended science curriculum of countries (Table 4.9). We undertook the same exercise at Grade 8 for the 1999, 2007 and 2015 TIMSS cycles; at Grade 8 all science topics appeared in the intended curricula of more than 50% of the participating countries (Table 4.10).

The growth in the number of Grade 4 TIMSS science topics taught in the majority of participating countries increased most in life sciences (from one to four). Earth sciences also saw a modest increase from zero to two topics in 2015. There was no change in the total number of physical science topics that were taught in over 80% of countries. Only one topic, "characteristics of living things and the major groups of living things", was taught in over 80% of countries across all three TIMSS cycles considered. This topic is fundamental to biology and so its presence as a core topic across years is perhaps unsurprising.

Table 4.9 Core and non-core TIMSS science topics at Grade 4

Life sciences	2003	2007	2015
Characteristics of living things and the major groups of living things (e.g. mammals, birds, insects, flowering plants)			
Major body structures and their functions in humans, other animals, and plants			
Life cycles of common plants and animals (e.g. humans, butterflies, frogs, flowering plants)			
Understanding that some characteristics are inherited and some are the result of the environment			
How physical features and behaviors help living things survive in their environments			
Relationships in communities and ecosystems (e.g. simple food chains, predator-prey relationships, human impacts on the environment)			
Human health (transmission and prevention of diseases, symptoms of health and illness, importance of a healthy diet and exercise)			
Physical sciences	2003	2007	2015
States of matter (solid, liquid, gas) and properties of the states of matter (volume, shape); how the state of matter changes by heating or cooling			
Classifying materials based on physical properties (e.g. weight / mass, volume, conducting heat, conducting electricity, magnetic attraction)			
Mixtures and how to separate a mixture into its components (e.g. sifting, filtering, evaporation, using a magnet)			
Chemical changes in everyday life (e.g., decaying, burning, rusting, cooking)			
Common sources of energy (e.g. the Sun, electricity, wind) and uses of energy (heating and cooling homes, providing light)			

	2003	2007	2015
Light and sound in everyday life (e.g. understanding shadows and reflection, understanding that vibrating objects make sound)			
Electricity and simple circuits (e.g. identifying materials that are conductors, recognizing that electricity can be changed to light or sound, knowing that a circuit must be complete to work correctly)			
Properties of magnets (e.g. knowing that like poles repel and opposite poles attract, recognizing that magnets can attract some objects)			
Forces that cause objects to move (e.g. gravity, pushing/pulling)			
Earth sciences	2003	2007	2015
Common features of the Earth's landscape (e.g. mountains, plains, deserts, rivers, oceans) and their relationship to human use (farming, irrigation, land development)			
Where water is found on the Earth and how it moves in and out of the air (e.g. evaporation, rainfall, cloud formation, dew formation)			
Understanding that weather can change from day to day, from season to season, and by geographic location			
Understanding what fossils are and what they can tell us about past conditions on Earth			
Objects in the solar system (the Sun, the Earth, the Moon, and other planets) and their movements (the Earth and other planets revolve around the Sun, the Moon revolves around the Earth)			

Key

Topic in curriculum of at least 80% of countries in TIMSS
Topic in curriculum of 50–80% of countries
Topic in curriculum of less than 50% of countries

There was only a small growth in the number of Grade 8 TIMSS science topics taught in the majority of participating countries' curricula (Table 4.8). The variation in the change across the three TIMSS cycles for each of the four content domains is therefore not unexpected: biology and earth science saw both an increase and decrease in the number of science topics taught in more than 80% of participating countries, while chemistry and physics saw an increase and no change respectively

Table 4.10 Core and non-core TIMSS science topics at Grade 8

Biology	1999	2007	2015
Differences among major taxonomic groups of organisms (plants, animals, fungi, mammals, birds, reptiles, fish, amphibians)			
Major organs and organ systems in humans and other organisms (structure / function, life processes that maintain stable bodily conditions)			
Cells, their structure and functions, including respiration and photosynthesis as cellular processes			
Life cycles, sexual reproduction, and heredity (passing on of traits, inherited versus acquired / learned characteristics)			
Role of variation and adaptation in survival / extinction of species in a changing environment (including fossil evidence for changes in life on Earth over time)			
Interdependence of populations of organisms in an ecosystem (e.g. energy flow, food webs, competition, predation) and factors affecting population size in an ecosystem			
Human health (causes of infectious diseases, methods of infection, prevention, immunity) and the importance of diet and exercise in maintaining health			
Chemistry	1999	2007	2015
Classification, composition, and particulate structure of matter (elements, compounds, mixtures, molecules, atoms, protons, neutrons, electrons)			
Mixtures and solutions (solvent, solute, concentration / dilution, effect of temperature on solubility)			
Properties and uses of common acids and bases			
Chemical change (transformation of reactants, evidence of chemical change, conservation of matter, common oxidation reactions–combustion, rusting, tarnishing)			

Physics	1999	2007	2015
Physical states and changes in matter (explanations of properties in terms of movement and distance between particles; phase change, thermal expansion, and changes in volume and / or pressure)			
Energy forms, transformations, heat, and temperature			
Basic properties / behaviors of light (reflection, refraction, light and color, simple ray diagrams) and sound (transmission through media, loudness, pitch, amplitude, frequency)			
Electric circuits (flow of current; types of circuits–parallel / series) and properties and uses of permanent magnets and electromagnets			
Forces and motion (types of forces, basic description of motion, effects of density and pressure)			

Earth science	1999	2007	2015
Earth's structure and physical features (Earth's crust, mantle, and core; composition and relative distribution of water, and composition of air)			
Earth's processes, cycles, and history (rock cycle; water cycle; weather versus climate; major geological events; formation of fossils and fossil fuels)			
Earth's resources, their use and conservation (e.g. renewable / non-renewable resources, human use of land / soil, water resources)			
Earth in the solar system and the universe (phenomena on Earth–day / night, tides, phases of moon, eclipses, seasons; physical features of Earth compared to other bodies)			

Key

Topic in curriculum of at least 80% of countries in TIMSS
Topic in curriculum of 50–80% of countries
Topic in curriculum of less than 50% of countries

in the number of topics taught in more than 80% of countries. No particular content domain is becoming increasingly more common across all countries.

Six science topics were taught in over 80% of participating countries across all three TIMSS cycles considered. "Cell structure and function" is one of these core topics and is again a topic that has been identified as one of the fundamental concepts in school science (Harlen 2010). It is thus unsurprising that it features as a core topic throughout.

4.1.6   Comparison of TIMSS Topics Taught Across Countries

We have identified science topics that are considered core and non-core in the science curriculum of countries participating in TIMSS (Sect. 4.1.5). However, as the number of countries that has participated in TIMSS has varied between cycles, comparisons are challenging. To overcome this, we focused on science topics included in the science curriculum of countries that have participated in all three TIMSS cycles under consideration for Grade 4 (Table 4.11) and for Grade 8 (Table 4.12).

At Grade 8 a far higher number of the TIMSS science topics were included in the curriculum of most participating countries than at Grade 4. At both grades there were a number of topics that were taught in the majority of cases across all TIMSS cycles, for example, "states of matter" at Grade 4 (Table 4.11) and "life cycles, sexual reproduction and heredity" at Grade 8 (Table 4.12). However, at Grade 4 in particular, there are some topics that are not commonly included in the curriculum in any of the three TIMSS cycles considered. One such topic is "understanding what fossils are and what they can tell us about past conditions on Earth".

We found that restricting the analysis to countries participating in all three TIMSS cycles under consideration nonetheless revealed very similar patterns to those we earlier identified in the wider pool of countries that had participated in at least one of the TIMSS cycles under investigation (see Sect. 4.1.4).

4.2   Cluster Analysis and Discriminant Analysis

We applied cluster and discriminant analyses of the curriculum questionnaire data to determine potential convergence of curricula; countries may be clustered into groups on the basis of the topics included or not included in their intended science curricula at Grade 4 and Grade 8.

4.2.1   Grade 4

The cluster analysis performed on the 2003 TIMSS data for Grade 4 resulted in countries being aggregated in two groups. As mentioned previously, the optimal number of groups is assessed as that with the lowest Bayesian information criterion

Table 4.11 Number of countries teaching Grade 4 TIMSS science topics in all three cycles

TIMSS science topic	Number of countries including topic in curriculum (maximum 18)*		
	2003	2007	2015
Characteristics of living things and the major groups of living things (e.g. mammals, birds, insects, flowering plants)	18	17	17
Major body structures and their functions in humans, other animals, and plants	17	17	16
Life cycles of common plants and animals (e.g. humans, butterflies, frogs, flowering plants)	16	15	18
Understanding that some characteristics are inherited and some are the result of the environment	11	9	5
How physical features and behaviors help living things survive in their environments	12	11	14
Relationships in communities and ecosystems (e.g. simple food chains, predator-prey relationships, human impacts on the environment)	13	14	12
Human health (transmission and prevention of diseases, symptoms of health and illness, importance of a healthy diet and exercise)	10	11	13
States of matter (solid, liquid, gas) and properties of the states of matter (volume, shape); how the state of matter changes by heating or cooling	17	16	17
Classifying materials based on physical properties (e.g. weight / mass, volume, conducting heat, conducting electricity, magnetic attraction)	17	16	11
Mixtures and how to separate a mixture into its components (e.g. sifting, filtering, evaporation, using a magnet)	9	8	7
Chemical changes in everyday life (e.g. decaying, burning, rusting, cooking)	8	9	8
Common sources of energy (e.g. the Sun, electricity, wind) and uses of energy (heating and cooling homes, providing light)	6	9	12
Light and sound in everyday life (e.g. understanding shadows and reflection, understanding that vibrating objects make sound)	11	16	14

Electricity and simple circuits (e.g. identifying materials that are conductors, recognizing that electricity can be changed to light or sound, knowing that a circuit must be complete to work correctly)	12	10	10
Properties of magnets (e.g. knowing that like poles repel and opposite poles attract, recognizing that magnets can attract some objects)	13	11	11
Forces that cause objects to move (e.g. gravity, pushing / pulling)	9	8	12
Common features of the Earth's landscape (e.g. mountains, plains, deserts, rivers, oceans) and their relationship to human use (farming, irrigation, land development)	13	12	9
Where water is found on the Earth and how it moves in and out of the air (e.g. evaporation, rainfall, cloud formation, dew formation)	12	15	14
Understanding that weather can change from day to day, from season to season, and by geographic location	9	12	16
Understanding what fossils are and what they can tell us about past conditions on Earth	7	8	5
Objects in the solar system (the Sun, the Earth, the Moon, and other planets) and their movements (the Earth and other planets revolve around the Sun, the Moon revolves around the Earth)	9	12	11

Note Eighteen countries participated in all three cycles considered here.

Key

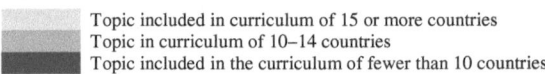

Topic included in curriculum of 15 or more countries
Topic in curriculum of 10–14 countries
Topic included in the curriculum of fewer than 10 countries

(BIC).[1] The cluster quality could be described as fair according to the silhouette measure of cohesion and separation.[2] Cluster analysis identified the first five most discriminating items for TIMSS 2003 at Grade 4 (Fig. 4.1).

---

[1]Hereafter we report the BIC for the solution with 1, 2 and 3 clusters:
 − 1 cluster: BIC = 1022.543
 − 2 clusters: BIC = 992.822 (−29.721 with respect to one single cluster)
 − 3 clusters: BIC = 1042.923 (+50.101 with respect to two clusters).

[2]This is a measure of how similar an object is to its own cluster (cohesion) compared to other clusters (separation). This measure ranges from −1 to 1, the highest figures indicating a better cluster outcome. In this case, the value of the silhouette measure is around 0.25 which falls into the 'fair' range (between 0.2 and 0.5).

Table 4.12 Number of countries teaching Grade 8 TIMSS science topics in all three cycles

TIMSS science topic	Number of countries including topic in curriculum (maximum 21)*		
	1999	2007	2015
Differences among major taxonomic groups of organisms (plants, animals, fungi, mammals, birds, reptiles, fish, amphibians)	21	21	18
Major organs and organ systems in humans and other organisms (structure / function, life processes that maintain stable bodily conditions)	18	21	20
Cells, their structure and functions, including respiration and photosynthesis as cellular processes	21	20	20
Life cycles, sexual reproduction, and heredity (passing on of traits, inherited versus acquired / learned characteristics)	21	21	20
Role of variation and adaptation in survival / extinction of species in a changing environment (including fossil evidence for changes in life on Earth over time)	14	12	16
Interdependence of populations of organisms in an ecosystem (e.g. energy flow, food webs, competition, predation) and factors affecting population size in an ecosystem	17	20	19
Human health (causes of infectious diseases, methods of infection, prevention, immunity) and the importance of diet and exercise in maintaining health	16	15	16
Classification, composition, and particulate structure of matter (elements, compounds, mixtures, molecules, atoms, protons, neutrons, electrons)	14	21	20
Mixtures and solutions (solvent, solute, concentration / dilution, effect of temperature on solubility)	20	17	19
Properties and uses of common acids and bases	19	17	15
Chemical change (transformation of reactants, evidence of chemical change, conservation of matter, common oxidation reactions–combustion, rusting, tarnishing)	14	16	17
Physical states and changes in matter (explanations of properties in terms of movement and distance between particles; phase change, thermal expansion, and changes in volume and/or pressure)	21	21	18
Energy forms, transformations, heat, and temperature	20	17	19
Basic properties / behaviors of light (reflection, refraction, light and color, simple ray diagrams) and sound (transmission through media, loudness, pitch, amplitude, frequency)	17	18	18
Electric circuits (flow of current; types of circuits–parallel / series) and properties and uses of permanent magnets and electromagnets	20	17	19
Forces and motion (types of forces, basic description of motion, effects of density and pressure)	16*	21	20
Earth's structure and physical features (Earth's crust, mantle, and core; composition and relative distribution of water, and composition of air)	17	19	20

Earth's processes, cycles, and history (rock cycle; water cycle; weather versus climate; major geological events; formation of fossils and fossil fuels)	15	21	19
Earth's resources, their use and conservation (e.g. renewable / non-renewable resources, human use of land / soil, water resources)	19	17	18
Earth in the solar system and the universe (phenomena on Earth–day / night, tides, phases of moon, eclipses, seasons; physical features of Earth compared to other bodies)	19	16	19

Note Twenty-one countries participated in all three cycles considered here.

Key

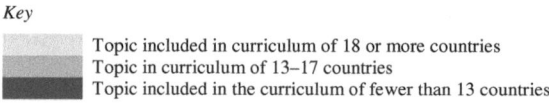

Topic included in curriculum of 18 or more countries
Topic in curriculum of 13–17 countries
Topic included in the curriculum of fewer than 13 countries

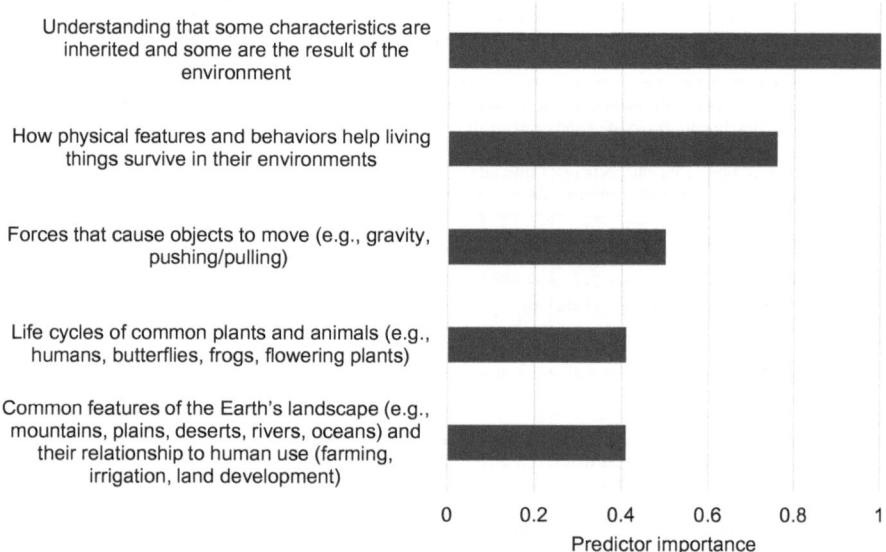

Fig. 4.1 Most discriminating topics in 2003 cluster analysis: Grade 4

The most discriminating science topic was "Understanding that some characteristics are inherited and some are the result of the environment", indicating that this specific science topic is the one that is the most important in defining whether a country may be classified as belonging to Group 1 or Group 2 (Fig. 4.2). For each of the five most discriminating science topics, we mapped the number of countries in Group 1 and Group 2 that were teaching the science topic to all or most students, only to the most able students, or not including the topic in the curriculum at that grade.

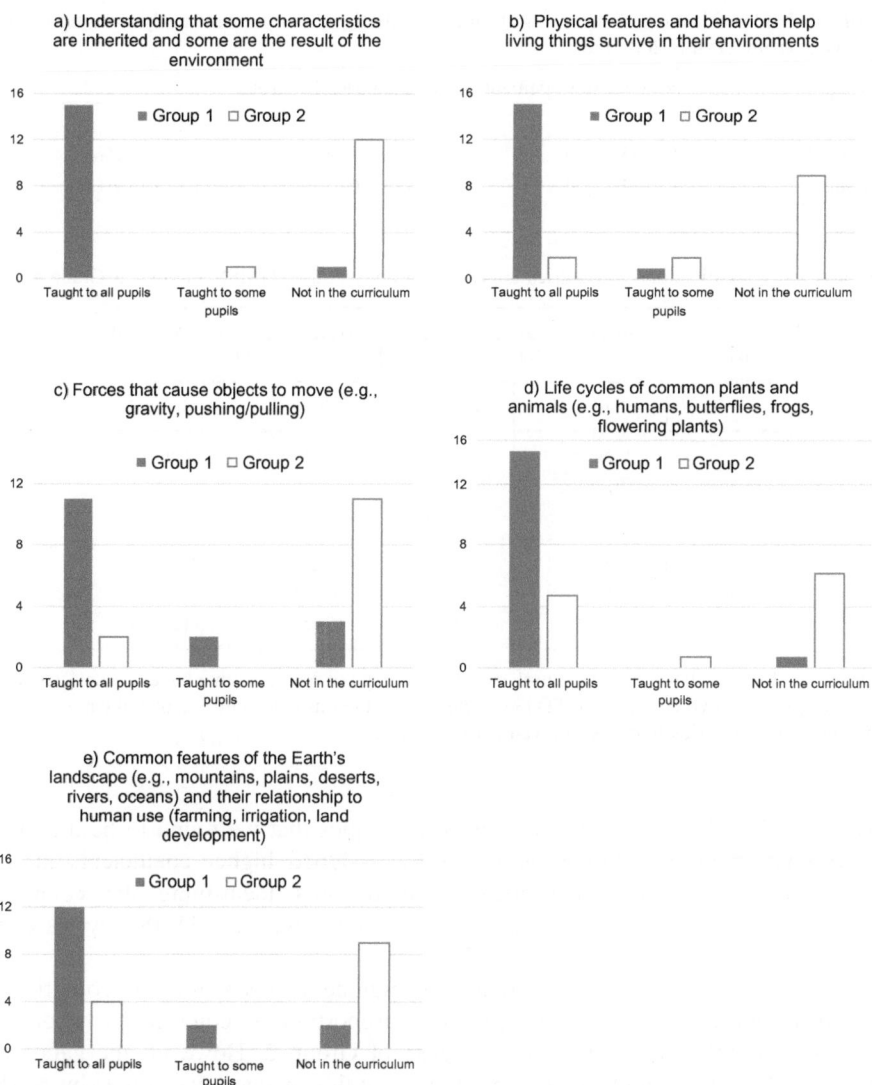

Fig. 4.2 Answers by group for the five most discriminating topics identified by cluster analysis of TIMSS 2003 data for Grade 4

While for the most discriminating item (Fig. 4.2a), there is an almost perfect separation between answers given by countries in Group 1 and countries in Group 2, the answers given by countries in the two groups tend to become more similar as the item becomes less discriminating (Fig. 4.2b–e).

Having identified two groups using cluster analysis, we conducted a discriminant analysis in order to estimate a classification model for countries in the successive

Table 4.13 Classification of countries at Grade 4 based on discriminant analysis using cluster groupings from TIMSS 2003 data

Group	Number (and percentage) of countries by group		
	2003	2007	2015
Group 1	16 (55.17%)	35 (76.09%)	37 (69.81%)
Group 2	13 (44.83%)	11 (23.91%)	16 (30.19%)

Table 4.14 Classification of countries appearing in consecutive TIMSS surveys at Grade 4

Initial group	Movement between 2003 and 2007		Movement between 2007 and 2015	
	Group 1	Group 2	Group 1	Group 2
Group 1	10 out of 11 remained (90.91%)	1 out of 11 moved (9.09%)	16 out of 21 remained (76.19%)	5 out of 21 moved (23.81%)
Group 2	5 out of 11 moved (45.45%)	6 out of 11 remained (54.55%)	5 out of 9 moved (55.56%)	4 out of 9 remained (44.44%)

Note Gray-shaded cells contain the number and percentage of countries that have been classified in the same group in two consecutive TIMSS cycles, while the unshaded cells display the number and percentage of countries that have moved across groups

cycles of TIMSS. As already mentioned, the topics that we found to be the most discriminating in the cluster analysis were assigned higher coefficients in the regression. These topics contribute most towards identifying the countries belonging to one of the two groups in the second and third TIMSS cycles considered (Tables 4.13 and 4.14).[3]

For each TIMSS cycle, we ascertained the number and percentage of countries in Group 1 and Group 2 (Table 4.13). The proportion of countries in Group 1 increased slightly over time at the expense of Group 2. However, the gain was modest (+14.64% points) and the possibility that this increase was driven by newly entered countries that did not participate in TIMSS 2003 cannot be ruled out.

Countries moved between groups in subsequent cycles of the survey (Table 4.14).[4] For instance, we can see that out of the 11 countries that were classified as belonging to Group 1 in 2003, ten were still classified as belonging to

---

[3]It must be emphasized that, while the classification of countries in 2003 was the result of our cluster analysis, the classification in the following two cycles was a prediction obtained by applying the model estimated from the discriminant analysis to the countries' new set of responses.

[4]By construction, these tables include only countries that have taken part in two consecutive TIMSS cycles.

Table 4.15 Cluster analysis classification of newly entered countries at Grade 4

Group	2007	2015
Group 1	19 out of 24	17 out of 23
	(79.17%)	(73.91%)
Group 2	5 out of 24	6 out of 23
	(20.83%)	(26.09%)

the same group in 2007, while the remaining country was classified as belonging to Group 2. Hence, between 2003 and 2007 countries in Group 1 were mostly stable. Conversely, a non-negligible fraction of countries belonging to Group 2 (45.45%) changed their classification. Between 2007 and 2015 few countries moved from Group 1 to Group 2 (23.81%) while, once again, a considerable fraction of countries that were classified as belonging to Group 2 (55.56%) moved to Group 1. This provides some evidence in support of the hypothesis that science curricula are converging and becoming more similar over time at Grade 4.

The critical point is that, while in the first table the increase in the percentage of countries belonging to Group 1 might be driven by new countries that were not included in previous cycles (see Table 4.13), the same set of countries are in each of the tables (see Table 4.14). Hence, movements (and as a consequence variations in the percentage of countries in each group) occur only when a country that was originally classified in one group modifies its science curriculum to such an extent that it becomes more similar to countries in the other group.

To conclude the investigation of the results obtained for Grade 4, consideration needs to be given to how newly entered countries[5] are classified in 2007 and 2015. The majority of countries that entered TIMSS in 2007 are classified as belonging to Group 1 and the same applies to countries that entered the study in 2015 (Table 4.15). The confluence of newly entered countries into one of the two groups does not indicate convergence by itself.[6] Nevertheless, if a core science curriculum is emerging we would expect the majority of countries to adhere to it.

The increase in the proportion of countries belonging to Group 1 (Tables 4.13 and 4.15) suggests that this is partially because new countries tend to be classified more often than not in this group. However, even considering only countries that have been in two consecutive TIMSS cycles, Group 1 is growing at the expense of Group 2, suggesting a certain degree of convergence (Table 4.14). The finding that newly entered countries are more often classified as belonging to Group 1 is consistent with the emergence of a core science curriculum to which countries tend to adhere.

For the five most discriminating items (see Fig. 4.1), we further analyzed the answers given by the 18 countries that participated in the 2003, 2007 and 2015

---

[5]These are countries that are in the dataset in a given cycle but not in the immediate previous cycle.
[6]Newly entered countries could be disproportionally more similar to countries in one of the two groups to begin with.

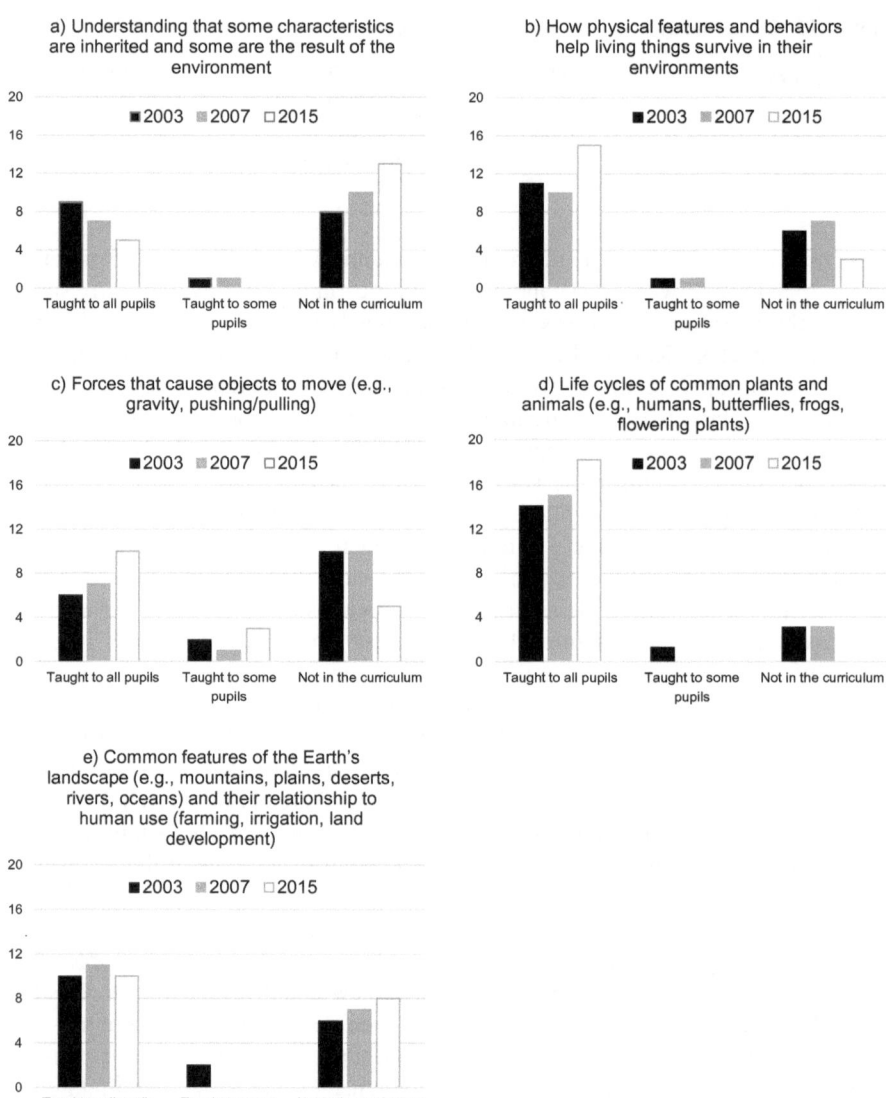

Fig. 4.3 Evolution of the distribution of the five most discriminating Grade 4 science topics (identified in 2003) between 2003 and 2015

TIMSS cycles (see Fig. 4.3a–e) to establish whether there were indications that science curricula at Grade 4 were converging. Only the 18 countries that participated in all the TIMSS cycles under consideration were included, to ensure statistical validity.

Three science topics demonstrated a clear tendency to converge[7] (Fig. 4.3a–c), while one remained stable (Fig. 4.3e) and the other underwent an inversion in tendency (the topic "forces that cause objects to move" was predominantly not taught in 2003, but the majority of countries had included it in the curriculum by 2015; see Fig. 4.3d). This is important because it confirms that the model we have presented identifies a general tendency to convergence in science curricula. The finding of overall convergence does not rule out the possibility that single topics are actually diverging. As long as the increase in similarities is enough to offset the emergence of dissimilarities, the model will identify overall convergence.

The model obtained by means of the discriminant analysis puts more emphasis on topics that were particularly important in defining the clusters in the first cycle and less on topics that were not discriminating at that point. We therefore repeated the same process in reverse. In practical terms, starting from the groups identified by means of a cluster analysis of the 2015 TIMSS cycle, we estimated a model that predicted countries' membership based on the responses they gave to the 2015 curriculum questionnaire. This model was then used to group countries in earlier cycles. We used this approach to verify whether aspects that were similar in 2003 (i.e. science topics that were widely taught or not taught) have become increasingly different over time. Our previous model was not able to identify this phenomenon, as topics that were not discriminating in the first cycle give only a very minor contribution in predicting future membership. The model once again identified two clusters[8] and the classification quality was fair according to the silhouette measure of cohesion and separation.

Among the five most discriminating items in 2015, there was only one topic ("understanding that some characteristics are inherited and some are the result of the environment") that was shared with the 2003 analysis (Fig. 4.4). This supported the hypothesis of convergence. The topics that were very discriminating between Group 1 and Group 2 in 2003 had converged and dissimilarities with respect to other topics now defined the clustering. These dissimilarities might be pre-existing (meaning that the topics were already dissimilar in 2003, but they were over-shadowed by other aspects that were even more relevant in defining the clustering of countries), or they might have emerged over time (topics that were widely taught, or not taught, by the majority of countries have been removed, or introduced, by some of them). The occurrence of this second event, which would indicate divergence along some dimensions, was why we repeated the analysis in reverse.

---

[7]In this case, convergence is identified by a concentration of the percentage reported in the histogram towards one single answer.

[8]The BIC for the solution with 1, 2 and 3 clusters:
  – 1 cluster: BIC = 1537.288
  – 2 clusters: BIC = 1450.742 (−86.546 with respect to one single cluster)
  – 3 clusters: BIC = 1494.038 (+43.296 with respect to two clusters).

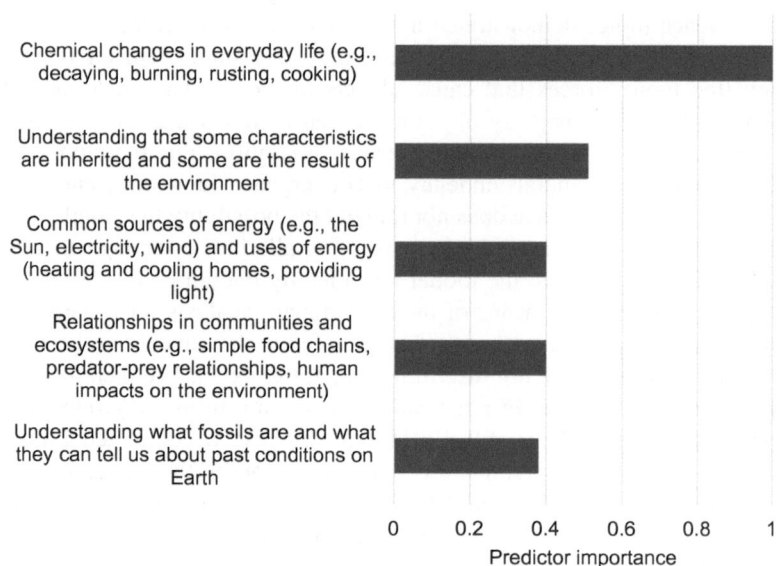

Fig. 4.4 Most discriminating items in 2015 cluster analysis: Grade 4

Table 4.16 Classification of countries at Grade 4 based on discriminant analysis using cluster groupings from TIMSS 2015 data

Group	Number (and percentage) of countries by group		
	2003	2007	2015
Group 1	22 out of 29	26 out of 46	31 out of 53
	(75.86%)	(56.52%)	(58.49%)
Group 2	7 out of 29	20 out of 46	22 out of 53
	(24.14%)	(43.48%)	(41.51%)

Applying this reverse approach, we found that the percentage of countries in Group 1[9] had decreased between 2003 and 2007 and remained stable in the following cycle (Table 4.16). This contradicts the hypothesis of convergence of science curricula, as it suggests that the treatment of topics that distinguished countries in 2015 was more homogeneous in 2003. Nevertheless, the results are based on different sets of countries in each cycle and hence the evidence might be driven by new entrants. To address this issue, we replicated our investigation into the movement of countries between consecutive cycles (Table 4.17). Once again, the general tendency was for countries to move from Group 2 to Group 1. However,

---

[9]Even though we call these two groups Group 1 and Group 2, these groups do not relate to the groups we have analyzed previously as the groups have been identified by means of two different models.

Table 4.17 Classification of countries appearing in consecutive TIMSS surveys at Grade 4

Initial group	Movement between 2003 and 2007		Movement between 2007 and 2015	
	Group 1	Group 2	Group 1	Group 2
Group 1	13 out of 17 remained (76.47%)	4 out of 17 moved (23.53%)	10 out of 16 remained (62.50%)	6 out of 16 moved (37.50%)
Group 2	3 out of 5 moved (60.00%)	2 out of 5 remained (40.00%)	9 out of 14 moved (64.29%)	5 out of 14 remained (35.71%)

Note Gray-shaded cells contain the number and percentage of countries that have been classified in the same group in two consecutive TIMSS cycles, while the unshaded cells display the number and percentage of countries that have moved across groups

Table 4.18 Cluster analysis classification of newly entered countries at Grade 4	Group	2007	2015
	Group 1	11 out of 24 (45.83%)	16 out of 23 (69.57%)
	Group 2	13 out of 24 (54.17%)	7 out of 23 (30.43%)

there was still a non-negligible fraction of countries that followed the reverse course, particularly between 2007 and 2015. Given that there appeared to be considerable movement across groups, the results provided only weak evidence in support of the hypothesis of convergence in science curricula (Table 4.17).

We also studied the classification of newly entered countries in 2007 and 2015 (Table 4.18).[10] The majority of countries that entered the survey in 2015 were classified as belonging to Group 1, while new entrants in 2007 were distributed almost equally between Group 1 and Group 2. The results are not conclusive for the reasons mentioned before, and there are no strong indications that a core science curriculum is emerging.

We interpreted these results (Tables 4.16, 4.17 and 4.18) as suggesting that topics that distinguished countries as belonging to Group 1 and Group 2 in 2015 were not more similar in 2003 (and thus there was no divergence).

---

[10]Even if the discriminant analysis is applied in reverse, the results of the model are interpreted in the same way as in the previous case. The only purpose for clustering countries according to questionnaire responses given in the 2015 TIMSS cycle is to give more importance to items that differ at this point in time. We are still interested in the effect of countries that entered the study in 2007 and 2015 on the percentages (Table 4.16).

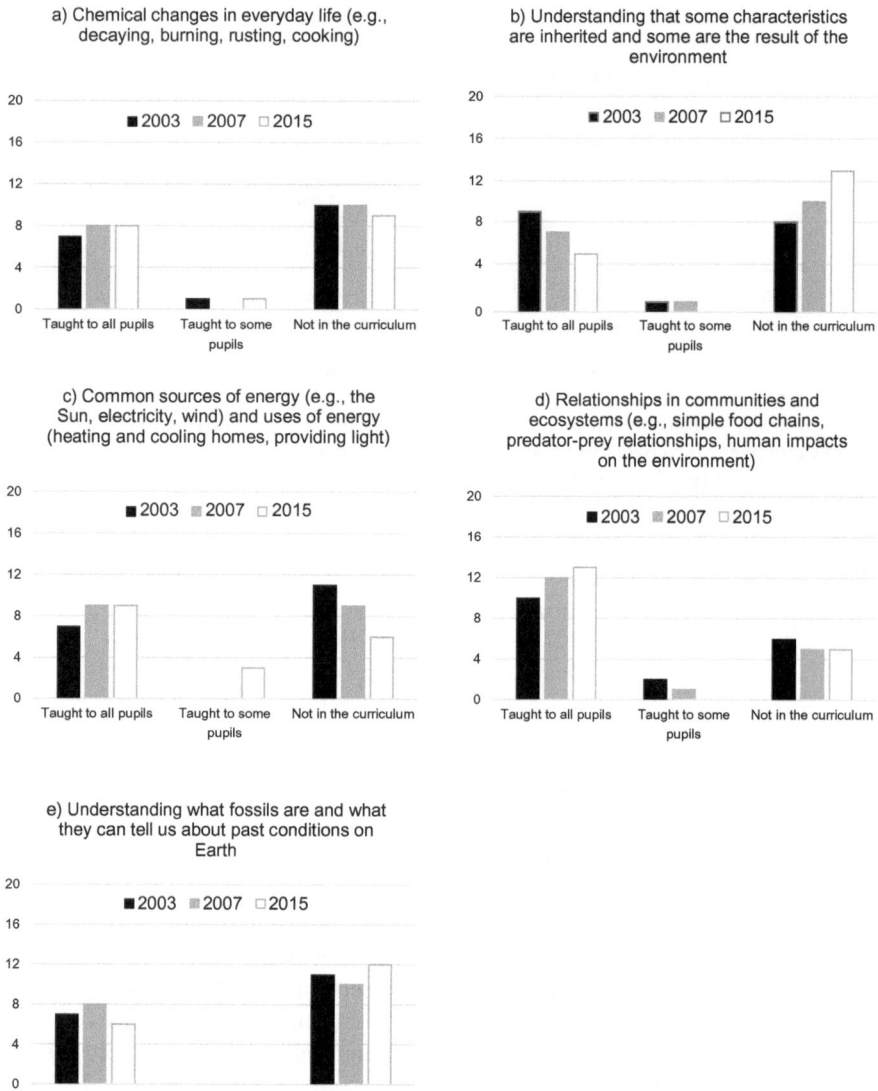

Fig. 4.5 Evolution of the distribution of the five most discriminating science topics (identified in 2015) between 2003 and 2015

We also examined how the distribution of the most discriminating items had evolved over time (Fig. 4.5). As in the previous case (Fig. 4.3), the behavior of topics was heterogeneous. Some were converging (e.g. see Fig. 4.5a, c), while for the others it was hard to identify a clear pattern.

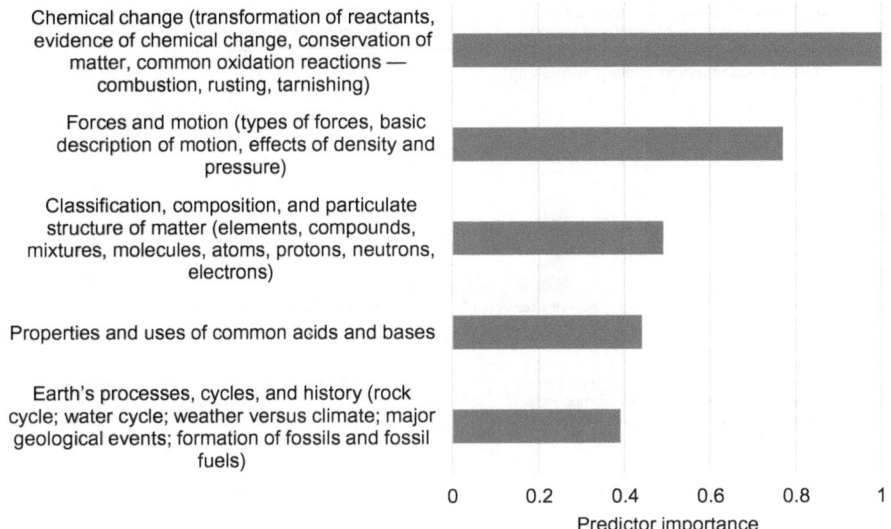

Fig. 4.6 Most discriminating science topics in TIMSS 1999 cluster analysis: Grade 8

Taking into account the outcomes presented so far, we conclude that there is only weak evidence to support the hypothesis that science curricula at Grade 4 have become increasingly similar over time.

4.2.2   Grade 8

We repeated the analysis we have presented for Grade 4 in order to assess convergence of science curricula at Grade 8.

As for Grade 4, the cluster analysis performed on the 1999 TIMSS data for Grade 8 resulted in countries being classified in two groups.[11] The cluster quality is fair according to the silhouette measure of cohesion and separation. Cluster analysis predicted the five most discriminating items for TIMSS 1999 (Fig. 4.6).

The most discriminating topic is "chemical change (transformation of reactants, evidence of chemical change, conservation of matter, common oxidation reactions–combustion, rusting, tarnishing)".

For each of the five most discriminating items (Fig. 4.7), we mapped the number of countries in Group 1 and Group 2 that were teaching the topic to all or most

---

[11]The BIC for the solution with 1, 2 and 3 clusters:

  – 1 cluster: BIC = 835.908
  – 2 clusters: BIC = 801.228 (−34.680 with respect to one single cluster)
  – 3 clusters: BIC = 838.015 (+36.787 with respect to two clusters).

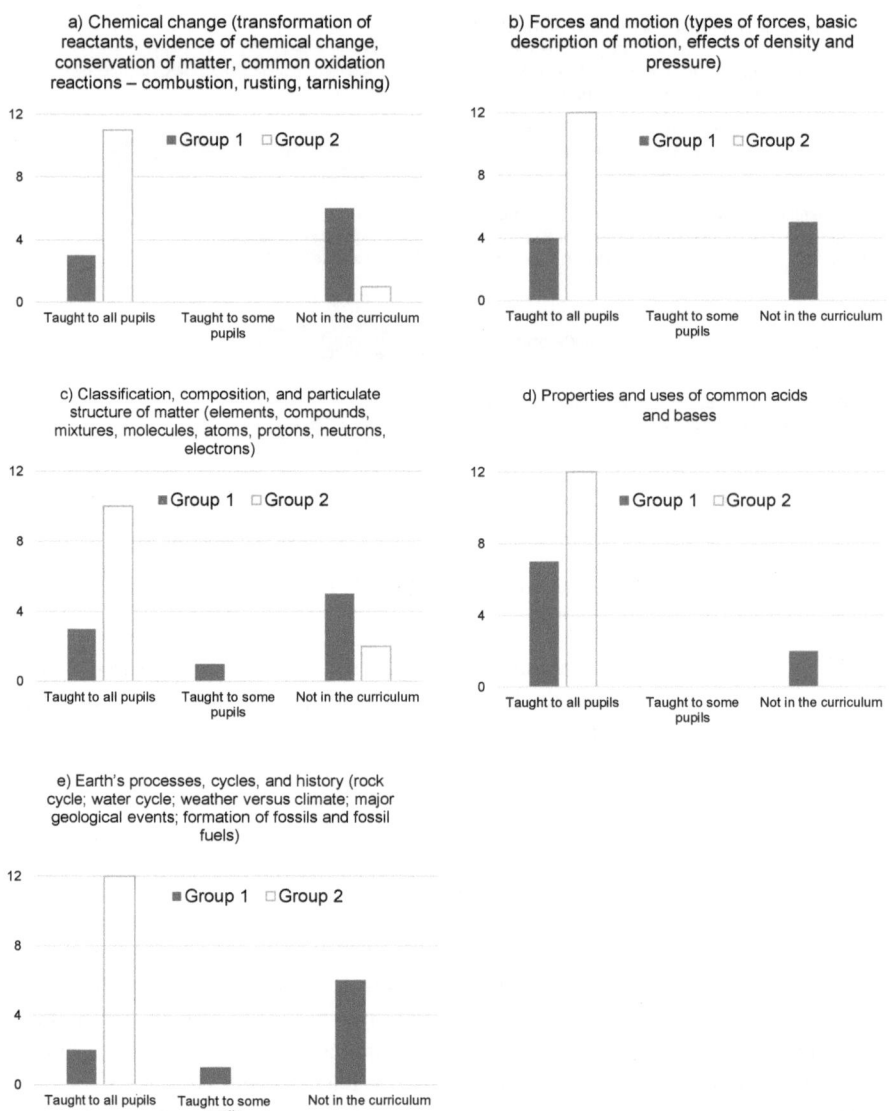

Fig. 4.7 Answers by group for the five most discriminating topics at Grade 8 identified by TIMSS 1999 cluster analysis

students, only the more able students, or not including the topic in the curriculum at that grade.

The more discriminating the item (from Fig. 4.7a–e), the sharper the distinction between Group 1 and Group 2.

Table 4.19 Classification of countries at Grade 8 based on discriminant analysis using cluster groupings from TIMSS 1999 data

Group	Number (and percentage) of countries by group		
	1999	2007	2015
Group 1	20 out of 38	9 out of 58	3 out of 45
	(52.63%)	(15.52%)	(6.67%)
Group 2	18 out of 38	49 out of 58	42 out of 45
	(47.37%)	(84.48%)	(93.33%)

Table 4.20 Classification of countries appearing in consecutive TIMSS surveys at Grade 8

	2007		2015	
Initial group	Group 1	Group 2	Group 1	Group 2
Group 1	3 out of 12 remained	9 out of 12 moved	0 out of 5 remained	5 out of 5 moved
	(25.00%)	(75.00%)	(0.00%)	(100.00%)
Group 2	0 out of 15 moved	15 out of 15 remained	3 out of 31 moved	28 out of 31 remained
	(0.0 0%)	(100.00%)	(9.68%)	(90.32%)

Note Gray-shaded cells contain the number and percentage of countries that have been classified in the same group in two consecutive TIMSS cycles, while the unshaded cells display the number and percentage of countries that have moved across groups

We used the model that results from applying discriminant analysis on the clusters identified in 1999 to predict the probability of belonging to Group 1 and Group 2 in the later TIMSS cycles (2007 and 2015).

For each cycle, we ascertained the number and percentage of countries in Group 1 and Group 2 (Table 4.19), revealing a strong tendency for countries to concentrate in a single group over time. The two groups were of similar size in 1999, but the overwhelming majority of countries were classified in Group 2 by 2015. As highlighted previously, even given the apparent strength of this result, it has to be interpreted with some caution because the increase of one group at the expense of the other could be driven by new countries entering TIMSS in 2007 and 2015.

Countries moved between groups in subsequent cycles of the survey (Table 4.20). Between 1999 and 2007, the majority of countries that were classified as belonging to Group 1 moved to Group 2 (75%) while no country followed the reverse path (from Group 2 to Group 1). Similarly, between 2007 and 2015, all the countries remaining in Group 1 moved to Group 2, while only a small number of countries moved from Group 2 to Group 1 (9.68% of countries classified as belonging to Group 2 in 2007).

Table 4.21 Cluster analysis classification of newly entered countries at Grade 8

Group	2007	2015
Group 1	6 out of 31	0 out of 9
	(19.35%)	(0.00%)
Group 2	25 out of 31	9 out of 9
	(80.65%)	(100.00%)

In both 2007 and 2015, the great majority of new entrant countries (all new entrants in 2015) were classified as belonging to Group 2 (Table 4.21). The results (Tables 4.19 and 4.20) provide strong evidence in support of the hypothesis of convergence in science curricula at Grade 8.

We repeated the exercise we performed for Grade 4 for the five most discriminating items identified by the 1999 cluster analysis, further analyzing the answers given by the 21 countries that participated in 1999, 2007 and 2015 (Fig. 4.8) to clarify why countries have been increasingly classified as belonging to Group 2 over time. Only data for the 21 countries that participated in all three TIMSS cycles under consideration were included, to ensure statistical validity.

In contrast to Grade 4, among the five most discriminating topics, all items but one ("properties and uses of common acids and bases") demonstrated a clear tendency to converge.

Repeating the cluster and discriminant analyses in reverse starting from the 2015 data to verify the findings once again identified two groups of countries,[12] with "fair" cluster quality according to the silhouette measure of cohesion and separation.

Among the five most discriminating topics (Fig 4.9), there was only one item ("Earth's processes, cycles, and history") that was common to the 1999 and 2015 analyses. This confirms the convergence we already identified; the topics that were very discriminating in 1999 converged and other topics were driving the clustering.

Applying this reverse approach, we identified the number and proportion of countries classified in Group 1 and Group 2 over the three cycles of TIMSS at Grade 8 (Table 4.22). The percentage of countries in both groups was stable over time.

We also studied the movement of countries that participated in consecutive TIMSS cycles in order to rule out the possibility that the results (Table 4.22) were driven by countries that entered later cycles (Table 4.23). While there was some evidence that countries classified as belonging to Group 1 had moved to Group 2 in the ensuing cycle, movements in the opposite direction (from Group 2 to Group 1) were rare, evidencing a tendency for countries to concentrate towards Group 2. As a result, considering only the countries that have been in subsequent TIMSS cycles,

---

[12]The BIC for the solution with 1, 2 and 3 clusters:
  – 1 cluster: BIC = 1123.414
  – 2 clusters: BIC = 1072.458 (−50.955 with respect to one single cluster)
  – 3 clusters: BIC = 1104.873 (+32.415 with respect to two clusters).

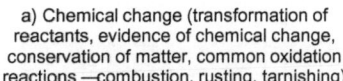

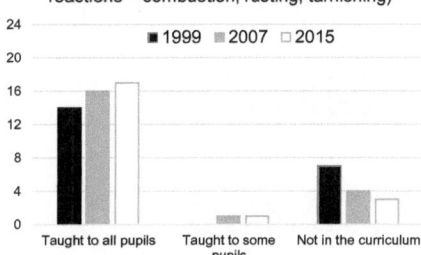

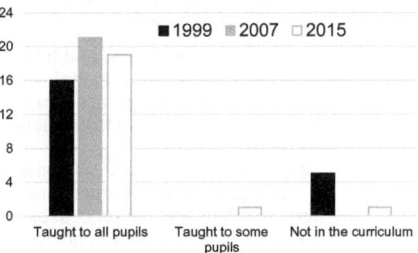

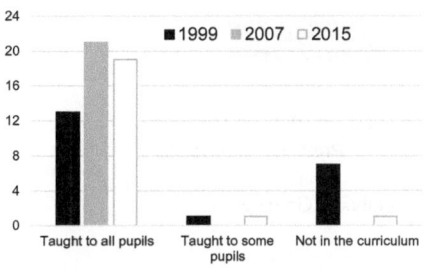

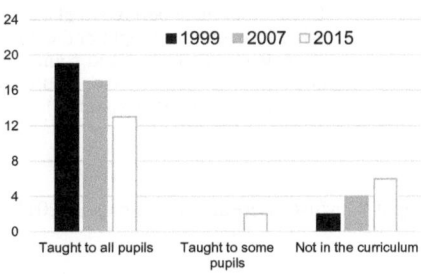

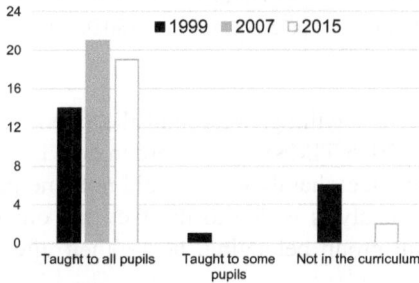

Fig. 4.8 Evolution of the distribution of the five most discriminating topics at Grade 8 (identified in 1999) between 1999 and 2015

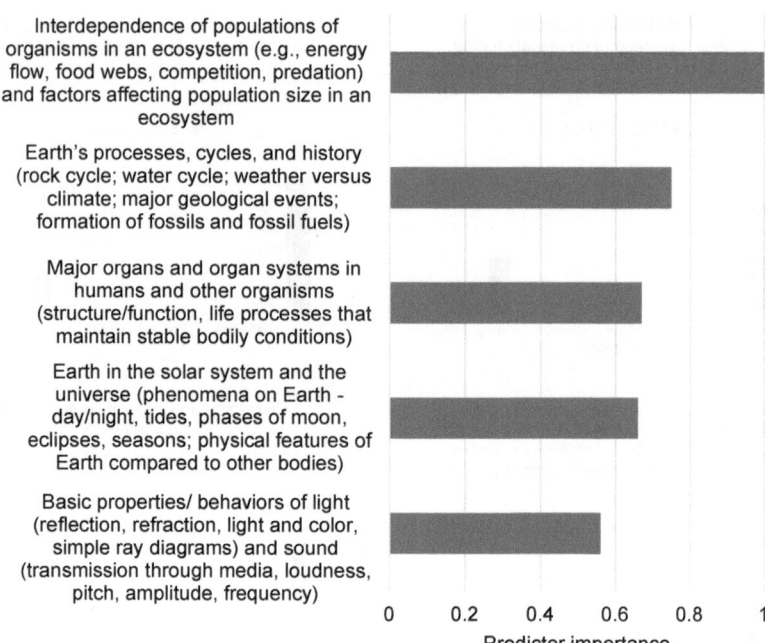

Fig. 4.9 Most discriminating topics in 2015 cluster analysis: Grade 8

Table 4.22 Classification of countries at Grade 8 based on discriminant analysis using cluster groupings from TIMSS 2015 data

Group	Number (and percentage) of countries by group		
	1999	2007	2015
Group 1	6 out of 38	14 out of 58	9 out of 45
	(15.79%)	(24.14%)	(20.00%)
Group 2	32 out of 38	44 out of 58	36 out of 45
	(84.21%)	(75.86%)	(80.00%)

not only did we find no sign of divergence, if anything, we found the opposite. Topics that were highly discriminating in the 2015 TIMSS cycle were treated in an even more unequal way in previous years. The fact that these items did not emerge as highly discriminating in the 1999 cluster analysis is due to the fact that other topics were even more important in defining group belonging, hence their effect was overshadowed.

Although confirming that the majority of countries entering TIMSS in 2007 and 2015 may be classified as belonging to Group 2 (Table 4.24) is not by itself conclusive, this does support the hypothesis that an internationally understood core science curriculum is developing over time.

Table 4.23 Classification of countries appearing in consecutive TIMSS surveys

	Movement between 2003 and 2007		Movement between 2007 and 2015	
Initial group	Group 1	Group 2	Group 1	Group 2
Group 1	2 out of 4 remained (50.00%)	2 out of 4 moved (50.00%)	3 out of 8 remained (62.50%)	5 out of 8 moved (37.50%)
Group 2	4 out of 23 moved (17.39%)	19 out of 23 remained (82.61%)	4 out of 28 moved (14.29%)	24 out of 28 remained (85.71%)

Note Gray-shaded cells contain the number and percentage of countries that have been classified in the same group in two consecutive TIMSS cycles, while the unshaded cells display the number and percentage of countries that have moved across groups

Table 4.24 Cluster analysis classification of newly entered countries

Group	2007	2015
Group 1	8 out of 31 (25.81%)	2 out of 9 (22.22%)
Group 2	23 out of 31 (74.19%)	7 out of 9 (77.78%)

Altogether, the results suggest that topics that differentiate countries belonging to Group 1 and Group 2 in 2015 were treated in an even more unequal way in previous cycles, indicating convergence (Tables 4.22, 4.23 and 4.24).

We also examined how the distribution of the most discriminating items (identified in 2015) evolved over time (Fig. 4.10). One topic ("Earth's processes, cycles, and history") displayed clear signs of convergence, while the remainder showed only minor or no evidence of convergence. This is mainly because there was already little variability in the 1999 TIMSS cycle, as science topics were treated in a similar way by all countries (the tendency was to teach these topics to all students in 1999). As a result, there is little capacity for further convergence. That said, we were mainly interested in confirming that no divergence occurred in topics that were not highly discriminating in the first cycle.

Taking into account the outcomes presented, we found strong evidence supporting the hypothesis of convergence in science curricula at Grade 8 over the last 20 years.

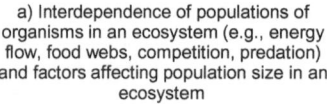

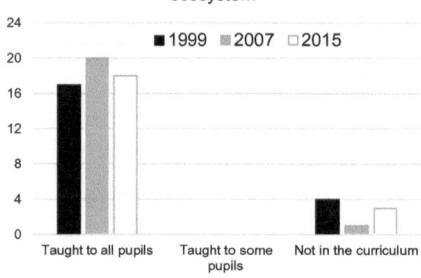

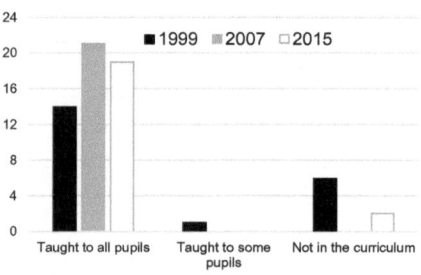

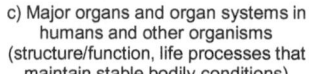

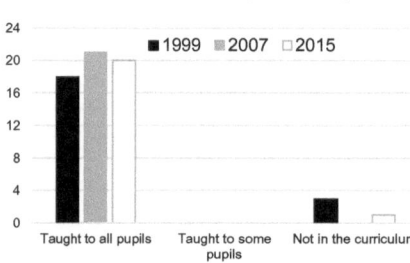

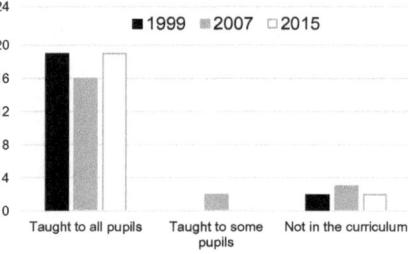

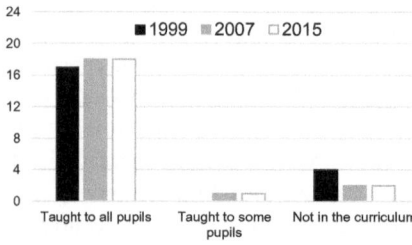

Fig. 4.10 Evolution of the distribution of the five most discriminating topics (identified in 2015) between 1999 and 2015

4.3 Further Analysis Using TIMSS Teacher Questionnaire and Encyclopedia Data

The results outlined in Sects. 4.1 and 4.2 are based on data from the TIMSS curriculum questionnaires. We also used the TIMSS encyclopedias, data from the TIMSS teacher questionnaire and additional data from the TIMSS curriculum questionnaire to build up a more detailed picture of other aspects of science curricula for a select number of countries (listed in Table 3.3). The objective of this analysis was to obtain information on aspects of countries' implemented science curricula to complement the data gathered on the intended curriculum (from the coding exercise and cluster and discriminant analyses). The aim was to enable changes in some aspects of the implemented science curricula to be tracked over time.

Curriculum features investigated included the instructional time spent teaching science each year and the percentage of students who have been taught the TIMSS science topics. Although we intended to explore a number of other curriculum features, the data sources did not support further additional analyses. Features that interested us included the emphasis teachers placed on science investigation in their teaching, the emphasis on different cognitive processes in the science curriculum, and the percentage of instructional time devoted to teaching science.

We encountered a number of difficulties in collecting information on the implemented science curricula. Firstly, for the 1999 (Grade 8) and 2003 (Grade 4) TIMSS cycles, there were no published TIMSS encyclopedias. The encyclopedias were only produced from the 2007 cycle onwards. As a result, much of the information on participating countries' science curricula that is available for later cycles was not available for the first TIMSS surveys considered in this report, thereby limiting the comparisons that can be made.

A second challenge was that, in many cases, the relevant items in the teacher and curriculum questionnaires changed between TIMSS cycles. This makes it impossible to directly compare responses between cycles. For example, in each TIMSS cycle there is a question that asks about teacher emphasis on science investigation; however, the wording of the question varies. In 2015, the question asked teachers to rate how often they emphasized science investigation in their lessons. They could select their response from "about half the lessons or more" or "in less than half the lessons". By contrast, in the 2007 teacher questionnaire, respondents were asked, for seven different science investigation activities, to rate whether the students they taught had the opportunity to do the activity in about half the lessons or more. These activities included practices such as watching the teacher demonstrate an investigation or relating what they learned about science to their daily lives. As the questions are framed in different ways, it is difficult to make direct comparisons. We encountered this issue for a large number of the curriculum features that we were originally intending to compare.

We also found that there was a considerable amount of missing data. We had selected countries for inclusion in this sample on the basis that they had taken part in at least two TIMSS cycles but, in a number of cases, countries had not provided this additional information on the implemented curriculum. For example, in 2007, there was no information provided by New Zealand on the percentage of students taught the TIMSS science topics for Grade 4 or Grade 8. Again, this made comparisons challenging, both for a single country over time and between countries in the same cycle. As a consequence, and because of the significant caveats that would have applied to any conclusions drawn from these analyses, we decided not to pursue further in-depth analyses based on the teacher and curriculum questionnaires.

Reference

Harlen, W. (Ed.). (2010). *Principles and big ideas of science education.* Hertfordshire: Association for Science Education. Retrieved from https://www.ase.org.uk/documents/principles-and-big-ideas-of-science-education/.

Chapter 5
The Globalization of Science Education and Science Curricula: Discussion and Conclusions

Abstract In this exploration of the globalization of science curricula, a coding exercise, cluster analysis and discriminant analysis of twenty years of TIMSS data were used to answer three research questions: (1) Have there been changes in intended science curricula over the last 20 years? (2) If changes do exist, do they support the hypothesis that science curricula are becoming increasingly similar across countries? (3) Are there groups of countries where curricula are increasingly similar; can the basis of an international core curriculum be identified? The analysis provides strong evidence to suggest that there have been changes to countries' intended science curricula. The coding exercise found that all countries in the analysis had made changes to their intended curricula, although the extent of curricular changes varied considerably between countries. Cluster and discriminant analyses showed that over time there was a tendency for countries to cluster into one particular grouping based on responses to the TIMSS curriculum questionnaire. The number of TIMSS science topics that could be considered core to the curricula of the majority of participating countries increased over time, particularly at Grade 4, suggesting that science curricula are becoming increasingly similar across countries. Among the two groups of countries identified by the cluster and discriminant analyses, there was a clear tendency for one group to include a wider science curriculum encompassing a greater range of science topics than the other group. At Grade 8 the results strongly suggest that there has been convergence in science curricula over time. In terms of whether an international core curriculum can be identified, there are some TIMSS science topics which could be considered core to the curricula of most countries. Assessment is likely to play an important role, as high-stakes assessment is likely to influence science curricula and what is taught in schools.

Keywords Core science curriculum · Curriculum change · Curriculum convergence · Curriculum development · Globalization · Science curriculum Science education · Trends in Mathematics and Science Study (TIMSS)

© International Association for the Evaluation of Educational Achievement (IEA) 2018   73
O. Stacey et al., *The Globalization of Science Curricula*, IEA Research
for Education 3, https://doi.org/10.1007/978-3-319-71532-2_5

5.1 Have There Been Changes in Intended Science Curricula Over the Last 20 Years?

Unsurprisingly, the intended science curriculum in many countries has clearly changed over the last 20 years due to multiple factors, including advances in scientific theory, changes in educational and assessment practices, changes to governmental emphases in education, as well as the influence of international large-scale assessments such as TIMSS and other globalizing factors (as discussed in Sect. 2.2). Our investigation therefore prompted additional but related supplementary questions:

- What are the extent and scope of changes to science curricula?
- Are there recognizable patterns to these changes that enable inferences to be drawn about which factors are important for different countries in influencing the specific nature of the changes that are made to their science curricula?

Before considering these questions, when drawing conclusions and making inferences from our analyses, note that the reference dataset we used was the TIMSS curriculum questionnaire. This questionnaire records, for each participating country, whether their science curriculum covered the individual topics within the TIMSS framework during the specified year of study. We identified science topics that mapped across all TIMSS cycles considered in this study; 20 science topics were defined for Grade 4 and 21 for Grade 8. The TIMSS questionnaire does not, however, record the topics included in a country's curriculum that are outside the TIMSS science framework, nor does it record information on topics that are present in the TIMSS science framework but taught to a year group other than the one specified in the framework. As a consequence, this is not a measure of the total science curriculum for each country, but a measure of the similarity of their intended science curriculum to the TIMSS framework.

5.1.1 The Scope and Extent of Changes to Science Curricula as Measured by the TIMSS Curriculum Questionnaire Datasets

In Sect. 4.1, we presented the overall extent of the changes that countries have made to their curricula, based on comparisons that can be made between countries that have recorded curriculum questionnaires for the two TIMSS cycles in question (Tables 4.5 and 4.6). From this we conclude that:

- For most countries, the majority of TIMSS science topics remained stable (i.e. their inclusion or omission from the countries' intended curricula did not change) across each pairwise comparison.

- TIMSS science topics were more likely to be added to national curricula than removed. This may indicate a trend towards alignment of national curricula with the TIMSS science framework, suggesting that, when introducing new topics to their curricula, countries may favor the TIMSS framework topics over other priorities.
- There were a higher number of changes evident at Grade 4 as compared to Grade 8. A possible explanation for this is that a much greater proportion of the Grade 8 topics were already included in participant countries' Grade 8 curricula than the equivalent proportion of Grade 4 topics. This suggests that there was greater "capacity for change" in the Grade 4 curricula. In addition, Grade 4 is much earlier in the learning development of individuals and therefore more distant from the pressures of the end-of-school qualifications that prepare students for entry into the (globalized) jobs market. This might mean that there are fewer external pressures for a standardized curriculum at Grade 4 than at Grade 8, allowing greater experimentation and therefore a more fluid curriculum.
- A small number of topics show a changed emphasis between TIMSS cycles for each grade. This is where topics may be introduced to a restricted range of students (usually the more able) or extended from a restricted range to the entire cohort. The indication is that this type of change is much less prevalent than adding or removing topics for the entire cohort, perhaps because of the additional logistical complications of defining and resourcing the delivery of a specific curriculum to subgroups within a cohort.

5.1.2 Patterns of Curricular Changes Made by Different Countries as Measured by the TIMSS Curriculum Questionnaire Datasets

As it is clear that there have been changes to the curricula of different countries across the TIMSS cycles, analyzing how these changes vary across different countries, and identifying any specific patterns in the changes that countries have introduced to their science curricula may provide important insights into the factors that may be driving such changes.

In order to investigate further, we focused our comparison of the curricula on countries that participated in the first available cycle of TIMSS (2003 for Grade 4 and 1999 for Grade 8) and the most recent cycle (2015 for both grades). We recorded all changes that had been made, including changes in the emphasis of a topic in the curriculum (namely changing a topic that had been specified as only for the most able students to a topic specified as appropriate for all students).

Accordingly, we grouped countries by the number of changes they made to their curricula with topics from the TIMSS framework (see Chap. 4, Tables 4.1 and 4.2). A greater proportion of countries made significant changes to their curriculum at Grade 4 than Grade 8, with 11 out of 20 countries at Grade 4 making changes to at least seven of the TIMSS framework topics (equivalent to a third of the framework topics). In contrast, only four of 24 countries at Grade 8 made a similar level of changes. This supports our earlier conclusion (Sect. 5.1.1) that changes are more common at Grade 4 than Grade 8.

Having identified the countries that made the greatest number of changes to their curricula between the specified time points, we investigated any apparent patterns that may explain why these changes were made. In our search for evidence of the influence of globalization effects, a particular focus was placed on whether the participation of countries in international large-scale assessments influenced the changes to their curricula, and specifically whether their participation and performance in TIMSS cycles had any influence on the changes they made. Robitaille et al. (2000) and Klieger (2015) both presented evidence that some countries do respond to their performance in international large-scale assessments with specific changes to their curricula.

The investigation focused on the countries that had made changes to at least a third of the relevant TIMSS science topics and compared their baseline performance in TIMSS, namely in 2003 (at Grade 4) or in 1999 (at Grade 8), to analyze the effect of changes at that grade. We used these baseline cycles as these provided the first internationally comparative data to governments of the performance of their students, and were therefore likely to have prompted discussion and review of the curriculum and how it may have been linked to performance in TIMSS. We also looked at the ranking of each of the countries in the TIMSS cycle to investigate whether performance in TIMSS could be related to the extent to which a country changed its science curriculum.

The IEA consistently emphasizes that the specific rankings in TIMSS are inaccurate, as a country's average scale score is presented as a range of values. The IEA also makes it clear that, when comparing performance, countries can only be judged as ranking separately if their scores are separated by at least the margin of error of measurement. However, within countries and in the international media, the nature of the ranking is often misunderstood and reported without such cautions, which may unduly influence policymakers into making changes to curricula.

Based on their TIMSS ranking, the 14 countries identified as having changed at least seven major topics of their curriculum divide into three distinct groups. These can be defined as:

- *Top-end competitors*: Countries that ranked between number two and number five in the rankings, but did not attain the top ranking, and so were perceived as competing for this in future cycles (Table 5.1).
- *Mid-ranking changers*: Countries that achieved scale scores above the international average but below the top five countries. These countries may be perceived as seeking to improve their position in the rankings (Table 5.2).
- *Below average responders*: Countries that achieved scores below the international average in the baseline cycle and have sought to respond with significant reform of their curriculum (Table 5.3).

Table 5.1 Top-end competitors

Grade	Country	Rank in baseline TIMSS	Number of changes	Type of changes	Baseline average scale score	2015 average scale score
4	Chinese Taipei	2/25	10	Adder	551	555
4	Japan	3/25	7	Adder	543	569[a]
4	Hong Kong	4/25	9	Balancer	542	557
8	Singapore	2/38	8	Cutter	568	597[a]
8	Japan	4/38	4	Adder	550	571[a]
8	Republic of Korea	5/38	6	Balancer	549	556

Key
Adder = a country where curriculum changes resulted in more TIMSS science topics appearing in the country's intended science curriculum
Balancer = A country where curriculum changes resulted in roughly equal number of TIMSS science topics being added and removed from the intended science curriculum
Cutter = A country where curriculum changes resulted in fewer TIMSS science topics appearing in the country's intended science curriculum
[a]Average scale score improved by at least 20 points from baseline to 2015 TIMSS. 20 points represents one-fifth of a standard deviation in TIMSS. An improvement in scale score was selected as opposed to an improvement in ranking as this is directly comparable across cycles

Table 5.2 Mid-ranking changers

Grade	Country	Rank in baseline TIMSS	Number of changes	Type of changes	Baseline average scale score	2015 average scale score
4	Australia	11/25	12	Cutter	521	524
4	New Zealand	12/25	8	Balancer	520	506
4	Belgium (Flemish)	13/25	15	Adder	518	512

Key
Adder = a country where curriculum changes resulted in more TIMSS science topics appearing in the country's intended science curriculum
Balancer = A country where curriculum changes resulted in roughly equal number of TIMSS science topics being added and removed from the intended science curriculum
Cutter = A country where curriculum changes resulted in fewer TIMSS science topics appearing in the country's intended science curriculum

Unexpectedly, we identified a significant number of changes to the curricula of countries with top-end performance. It could be expected that high-achieving countries would have a curriculum enabling high performance in TIMSS and would therefore be less likely to require multiple curriculum changes. A clue to the force

Table 5.3 Below average responders

Grade	Country	Rank in baseline TIMSS	Number of changes	Number of topics added	Baseline average scale score	2015 average scale score
4	Cyprus	19/25	7	Adder	480	481
4	Norway	20/25	10	Cutter	466	538[a]
4	Armenia	21/25	7	Adder	437	–
4	Morocco	25/25	13	Adder	304	352[a]
8	Israel	26/38	11	Adder	468	507[a]
8	Morocco	37/38	10	Balancer	323	393[a]
8	South Africa	38/38	12	Adder	243	358[a]

Key

Adder = a country where curriculum changes resulted in more TIMSS science topics appearing in the country's intended science curriculum

Balancer = A country where curriculum changes resulted in roughly equal number of TIMSS science topics being added and removed from the intended science curriculum

Cutter = A country where curriculum changes resulted in fewer TIMSS science topics appearing in the country's intended science curriculum

[a]Average scale score improved by at least 20 points from baseline to 2015 TIMSS. 20 points represents one-fifth of a standard deviation in TIMSS. An improvement in scale score was selected as opposed to an improvement in ranking as this is directly comparable across cycles

for change is that the countries that achieved the top rankings, Singapore at Grade 4 and Chinese Taipei at Grade 8, instituted only minimal change following their high performance at that grade. It is therefore of interest that these same countries instituted significant changes in the alternate grades where they both achieved second place. One possible reason is that performing just below the top spot has prompted a review of the curriculum in order to be more competitive in future TIMSS cycles. All countries that fall into this group are also geographically close. This might reflect a competitive effect amongst neighbors, or may indicate specific regional cultural influences.

The response in terms of the nature of curriculum change varied. The majority response was to add TIMSS science topics to the curriculum. Singapore, however, removed or reduced the emphasis of eight TIMSS topics in its Grade 8 curriculum (which is in contrast to its treatment of the Grade 4 curriculum, as discussed). Hong Kong and the Republic of Korea rebalanced their curricula by adding and removing TIMSS topics. However, the changes made to the topics taught in each country are dependent on the nature of the "baseline" curriculum for that country, so conclusions about the preferred approach are difficult to make.

As we were also interested in whether any of the specific changes were connected to the performance of a country in future TIMSS cycles, we also recorded the average scaled score for these countries in the baseline and 2015 cycles. Although all of these countries maintained their high ranking in the 2015 cycle, only Singapore at Grade 8 and Japan at both Grade 4 and Grade 8 improved their scale score by a noteworthy amount (more than 20 points) over this period. It is not possible to say which specific factors contributed to improvement in these scores, but it is likely that curriculum effects played an important part.

Another cluster of countries grouped in the middle of the ranking table made multiple curricular changes at Grade 4 but not at Grade 8 (Table 5.2). The most striking aspects of this group are that two of the countries changed more than half of the TIMSS science topics in their curriculum, but that, despite these large-scale changes, there was minimal impact on the average scaled score. For example, only Australia achieved a slightly higher score in TIMSS 2015 than the baseline, with New Zealand and Belgium achieving a lower score in the later cycle.

The "below average responders" group is where we expected curriculum change to be most apparent (Table 5.3). We assumed that countries might be likely to reflect on poor performance in the baseline TIMSS survey and use this as an impetus to change policy, whilst using the TIMSS framework as a model for modifying their national curricula. This is the process that Klieger (2015) detailed for Israel. In line with this hypothesis, the majority of "below average responders" have made a significant number of changes to their curriculum and added TIMSS content. The exceptions to this case were Norway, where the curriculum changes resulted in the removal of nine TIMSS science topics from its curriculum, and Morocco, where the curriculum changes resulted in an equal number of TIMSS science topics being added and removed from its science curriculum at Grade 8.

In terms of the impact of these changes, the "below average responders" have shown significant improvement in their TIMSS average scale scores, with Norway improving by 72 points at Grade 4 and moving to a position well above the international average. Israel also moved to above the international average at Grade 8 in 2015. That said, both Norway and Israel achieved scores close to the international average in their baseline year. Morocco (Grade 4 and Grade 8) and South Africa (Grade 8) also saw significant improvements in average scale score, with South Africa improving by 115 points over the time period. That said, this improvement was based on a poor baseline score and ranking, and South Africa remained the lowest ranked country in the 2015 TIMSS cycle. Morocco similarly only gained one and two ranking places at Grade 4 and Grade 8, respectively.

Although these findings suggest that closer adherence to the science curriculum represented by the TIMSS framework has boosted performance for the poorest performing countries in the TIMSS international assessment, causality cannot be inferred. A wide range of other potential variables, which have not been examined in this investigation, could contribute to this improvement. That said, improvement would be expected as the curriculum and assessment became much better aligned with the TIMSS framework. The analysis also suggests that countries value the improvements that this alignment can provide and that they see the TIMSS framework as a suitable model on which to base their curriculum development.

5.2   If Changes Do Exist, Do They Support the Hypothesis That Science Curricula Are Becoming Increasingly Similar Across Countries?

Addressing this research question was contingent on countries' intended science curricula changing over the past 20 years. Our evidence (Sect. 5.1) revealed that this has been the case for many countries, so the next step was to determine whether these changes in intended science curricula supported the hypothesis that science curricula were becoming increasingly similar across countries. Our analyses provided mixed evidence to support this hypothesis.

5.2.1   Evidence That Science Curricula Are Becoming Increasingly Similar

Our analysis suggests that, in some respects, the science curricula of countries participating in TIMSS are becoming increasingly similar. Firstly, the coding of curriculum questionnaires indicates that over time, at both Grade 4 and Grade 8, the number of TIMSS science topics that were included in countries' intended science curricula was increasing, with countries on average adding more topics to their science curricula than they are removing (see Chap. 4, Tables 4.5 and 4.6).

One consequence of this was that a greater number of the TIMSS science topics were included in the majority (over 80%) of participating countries' science curricula in the 2015 TIMSS cycle than in earlier cycles. The finding is consistent for both Grade 4 and Grade 8. At Grade 4, seven of the 21 TIMSS science topics were covered by more than 80% of participating countries in 2015, compared to only two of the TIMSS science topics in 2003. The trend was less pronounced at Grade 8, with an increase from 12 to 14 of the 20 TIMSS science topics included in over 80% of participating countries' curricula. In Grade 4, in addition, there was also an associated drop in the number of TIMSS science topics that only appeared in the minority (less than 50%) of participating countries' curricula. This trend was not observed at Grade 8 as, in the baseline (1999) TIMSS survey, there were no TIMSS science topics that appeared in less than 50% of participating countries' curricula.

These results provide evidence that countries' curricula are becoming more similar over time, and that these changes appear to be increasing the likelihood of the same science topics being included in each country's curriculum. When interpreting the outcomes of this analysis, it is important to acknowledge that the number and composition of countries in each of the TIMSS cycles varies and that this has the potential to impact the results. It is also important to acknowledge that the finding that science curricula are becoming increasingly similar relates only to the curriculum as measured against the science topics included in the TIMSS curriculum questionnaire. Whilst the science topics included in the TIMSS framework are broad and balanced in terms of coverage of different aspects of science, they are not an exhaustive list of the potential science topics that could be included in a country's curriculum. Additionally, we have no information regarding the science curricula of countries that have not participated in TIMSS and so cannot make any claims about the extent to which non-participating countries' science curricula are becoming more similar over time.

The outcomes of the cluster analysis and discriminant analysis provide some evidence to support the conclusion that science curricula are becoming more similar across countries participating in TIMSS, with the evidence being stronger at Grade 8 than at Grade 4. The analyses provide strong evidence of convergence in the science curricula of countries participating at Grade 8 between 1999 and 2015, with the number of countries clustering into Group 2 growing at the expense of the number clustering into Group 1. In 1999, approximately half of countries clustered into Group 1 and half into Group 2. By contrast, in 2015, over 90% of countries clustered into Group 2 and less than 10% into Group 1 (see Chap. 4, Table 4.19). This suggests that, over time, at Grade 8, changes in intended science curricula are leading to an increasing number of countries with similar profiles in terms of the topics they do and do not include in their science curricula.

At Grade 4, however, the strength of evidence from the cluster analysis and discriminant analysis is less strong. The discriminant analysis suggests that changes in participating countries' science curricula have led to only mild convergence over time. In 2003, 55% of countries clustered into Group 1 and 45% into Group 2, whereas in 2015, 70% of countries clustered into Group 1 and 30% into Group 2 (see Chap. 4, Table 4.13). However, there was some movement of countries between the two groups across the TIMSS cycles considered, suggesting less convergence and more fluidity over time than at Grade 8 (see Chap. 4, Table 4.14). This may be because the science curriculum at primary level in some countries is less prescribed than at secondary level, where the subject usually has a higher status and profile and is often associated with high-stakes assessment.

One additional piece of evidence from the cluster and discriminant analysis, which supports the hypothesis that science curricula are becoming increasingly similar over time, comes from the science topics that were most important in predicting which group a country clustered into. For example, in 1999 at Grade 8 the science topic that was most important at determining which group a country clustered into was "chemical change (transformation of reactants, evidence of chemical change, conservation of matter, common oxidation reactions–combustion, rusting, tarnishing)". By contrast, in the 2015 TIMSS cycle, this was only the fifteenth most important topic in determining which group a country clustered in. Of the 21 countries participating in all three TIMSS cycles considered in this investigation, in 1999, this topic was included in 11 out of the 12 Group 2 countries' intended science curriculum. By contrast, it was only included in three of the nine Group 1 countries' curriculum (Chap. 4, Fig. 4.5). However, in 2015, the majority of countries in both groups (17 out of 21) included this topic in their intended curriculum for all students (Chap. 4, Fig. 4.6).

The findings from our coding exercise and cluster and discriminant analyses are consistent with findings from the literature review. Robitaille et al. (2000), for example, found that, in response to the 1995 TIMSS cycle, a number of countries, such as Iran and Kuwait, changed the content of their science curricula to incorporate new science topics on the environment. These changes, in turn, led to their science curricula becoming more aligned to the TIMSS framework. Similarly, revisions to the Israeli science curriculum have led to closer alignment between this and the TIMSS science framework (Klieger 2015). As countries' science curricula change and become more aligned to the TIMSS framework, one consequence is that the intended science curricula of different countries will converge.

5.2.2   Evidence That Suggests Science Curricula Are Not Becoming Increasingly Similar

Although some aspects of the analyses and literature review supported the hypothesis that science curricula are becoming increasingly similar across countries, other aspects of the analysis suggested otherwise. Firstly, as already discussed, based on the outcomes of the cluster and discriminant analyses, the evidence for convergence in science curricula at Grade 4 was weaker than at Grade 8. This highlights the important point that it is too simplistic to make a broad generalization that science curricula are becoming increasingly similar across countries over time. The number of countries, phase of schooling (e.g. primary versus secondary) and time scale considered, all affect the conclusions that can be drawn with regard to this hypothesis.

Secondly, the majority of our analysis has focused on intended science curricula and, in other aspects of the curriculum, there is little evidence to support the notion that science curricula are becoming increasingly similar over time. For example, in terms of the amount of time spent teaching science, wide variations exist in the mean number of hours identified in each country, with no evidence of this converging over time. For example, in 2015 at Grade 4, the average time spent teaching science each year (as identified from the TIMSS teacher questionnaire) in Qatar was almost three times as much as in New Zealand. At Grade 8 a similar pattern emerged.

In addition, although the responses to the curriculum questionnaire indicated a tendency for more TIMSS science topics to be included in countries' intended science curricula in both grades over time, the teacher questionnaire responses painted a slightly different picture. For example, although we have not reported in detail on the results of our additional analysis of selected countries for the reasons cited in Sect. 4.4, we did find that the average percentage of students who had been taught the TIMSS science topics was slighter higher for both Grade 4 and Grade 8 in the baseline cycles (1999 for Grade 8, and 2003 for Grade 4) than in the most recent 2015 cycle. This suggests that, although there may be evidence to support the hypothesis of the *intended* science curriculum becoming increasingly similar over time, this is not necessarily reflected in the *implemented* science curriculum experienced by students. As research identified in the literature review suggested, although curricula can be centrally prescribed, the way in which they are interpreted and enacted can vary widely at the local level (Astiz et al. 2002; Cogan et al. 2001).

The literature review and outcomes of the discriminant analysis also suggest that, although science curricula may, in some respects, be becoming increasingly similar over time, the changes are not necessarily leading to uniformity, as countries still cluster into distinct groups. For both grades and in all cycles, our discriminant analysis produced two groupings. Other studies using a similar approach also identified multiple groupings, and this is true for both science (Kjaernsli and Lie 2008) and mathematics (Zanini and Benton 2015). These studies found a tendency for countries to group based on geographic and cultural lines, suggesting that these factors still play an important role in aspects of the curriculum for both science and mathematics.

5.3 Are There Groups of Countries Where Curricula Are Increasingly Similar; Can the Basis of an International Core Curriculum Be Identified?

We investigated whether there was any evidence that science curricula were becoming increasingly similar over time in certain groups of countries to establish whether the basis of an international core science curriculum could be identified.

5.3.1 Are There Groups of Countries Where Science Curricula Are Increasingly Similar?

We have already discussed the outcomes of the cluster and discriminant analyses (Sect. 5.2), which provide some evidence to suggest increased similarity of countries' science curricula over time.

Consideration of the science topics that most contribute to the country groupings enables the curricula of countries within each group to be categorized. Using the TIMSS 2015 data and the cluster and discriminant analysis for this year at both Grade 4 and Grade 8, it was possible to identify and characterize the curricula of the two groups of countries.

At Grade 4 in 2015, the main differences between the science curriculum profiles of Group 1 and Group 2 countries could be categorized by Group 1 countries tending to cover a greater number of science topics than Group 2 countries. The following topics, for example, were likely to be included in Group 1 countries' science curricula and unlikely to be present in Group 2 countries' science curricula:

- common sources of energy and uses of energy
- understanding that some characteristics are inherited and some are the result of the environment
- chemical changes in everyday life
- understanding what fossils are and what they can tell us.

Several of these topics come from the physical sciences area of the TIMSS science framework. This suggests that countries in Group 1 may have a more balanced and broad focus to their science curriculum. Countries in Group 2 appear to have less of a focus on physical science topics and instead focus on life sciences topics. It may be that physical science topics receive less attention as they can be more abstract and less concrete than some of the topics in life and Earth science, making them less likely to be taught at Grade 4. In addition, the life sciences topic "understanding that some characteristics are inherited and some are the result of the environment" is a relatively advanced concept for Grade 4 children and so it is perhaps unsurprising that it is not included in the curricula of many countries in Group 2.

A similar pattern emerged at Grade 8 in 2015, where Group 2 countries tended to include a greater number of science topics in their curriculum than Group 1 countries. The following four topics for example were likely to be included in Group 2 countries' science curricula and unlikely to be present in that of Group 1 countries:

- interdependence of populations of organisms in an ecosystem
- major organs and organ systems in humans and other organisms
- Earth's processes, cycles, and history
- Earth in the solar system and the universe.

In contrast to the Grade 4 topics, none of these topics were in the physical sciences areas of the curriculum (chemistry and physics). It is not so obvious why these topics may differ so much between the two groups, particularly as the two biology topics cover quite different aspects of the subject, one ecological and one covering human biology. It may be that the two Earth science topics, particularly Earth's processes, cycles, and history, may be covered by Group 2 countries in subjects other than science, for example in geography. This may explain the importance of this topic in the clustering, although this cannot be said for certain.

This clustering pattern and the finding that there are distinct groups of countries with similar curricula is consistent with other research, such as that conducted by Kjaernsli and Lie (2008), who investigated the achieved curriculum in different countries using item responses to TIMSS science questions for TIMSS 2003 data. This study also identified a number of distinct groupings of countries that clustered together. Similar patterns have emerged on analysis of TIMSS mathematics data (Rutkowski and Rutkowski 2009; Zanini and Benton 2015).

Where our data differs from these studies, however, is that the cluster groupings in previous studies have tended to identify a larger number of country groupings, whereas this analysis has identified only two. Additionally, the cluster groupings in previous studies have typically grouped along geographic or linguistic lines, for example an Arabic grouping, an East Asian grouping, an Anglophone grouping etc. With only two groups identified in our study for each cycle, and therefore a larger number of countries present in each group, it is more difficult to make generalizations about the geographic, cultural and linguistic characteristics of countries. For example, at Grade 4 in 2015, Anglophone and Scandinavian countries are present in both groups.

At Grade 8 in 2015, the proportion of countries in Group 2 is greater than in Group 1, which makes characterization of this group easier. Group 1 did not tend to contain Anglophone countries and tended to have a higher number of North African and Middle Eastern countries, including Morocco, Lebanon, Dubai and Bahrain. In addition, no Western European countries were present, suggesting that at this grade there was some country clustering along geographic and cultural lines.

Although the cluster and discriminant analysis provides evidence to suggest that there are distinct cluster groupings and that there are defined curriculum characteristics of countries in each cluster, it is important to note that there is fluidity in these groupings. For example, at Grade 4 in particular, there is movement of countries between groups between each TIMSS cycle. This suggests that, while there is increasing similarity in countries' curricula, countries still have the potential to change and shift their curriculum profile and characteristics between groupings.

5.3.2   Can the Basis of an International Core Science Curriculum Be Identified?

In addition to identifying country groupings where there is convergence of science curricula, the second element of this research question pertains to whether the basis of an international core curriculum can be identified. This research provides evidence to suggest that the basis of a core international science curriculum can be identified. For example, the coding exercise has identified TIMSS science topics which could now be described as core (included in over 80% of participating countries' curricula) at both Grade 4 (Table 4.9) and Grade 8 (Table 4.10).

At Grade 4, the number of TIMSS science topics taught in the majority of participating countries increased over time from two in 2003 to seven in 2015, out of a total of 21 topics (Table 4.7). At Grade 8, there was a more modest increase, from 12 to 14, out of a total of 20 topics (Table 4.8). This growth in the number of topics taught in the majority of participating countries, particularly at Grade 4, suggests that a core science curriculum is emerging amongst countries. The seven "core" topics at Grade 4 in 2015 consisted of four life sciences topics, one physical science and two Earth science topics. This is consistent with findings from the cluster and discriminant analysis where Group 2 countries tended not to teach some of the physical sciences topics, i.e. it is unsurprising that fewer physical sciences topics are represented in the core science topics at Grade 4.

Two of the topics which increased most in terms of the proportion of countries including them in their science curricula at Grade 4 were "human health" (which had an increase of 29 percentage points between 2003 and 2015), and "understanding that weather can change from day to day, from season to season, and by geographic location" (which had an increase of 27 percentage points). These topics reflect major global issues facing the world today. For example, the importance of a healthy diet and exercise can be viewed in light of rising levels of obesity which are affecting many developed countries across the world. Additionally, an understanding of weather patterns and variations is crucial to an appreciation of climate change.

At Grade 8 the core science topics were spread more evenly across the different science subjects. In TIMSS 2015, four biology topics, three chemistry topics, three physics topics and four Earth science topics could be considered core. This may reflect the fact that, at secondary level, as students are introduced to topics such as the particle model and the structure of atoms, more aspects of chemistry and physics become accessible to students and can be more readily included in science curricula.

The presence of an international core science curriculum and the evidence from this study which suggests that this core curriculum appears to be growing over time is consistent with research highlighted in the literature review identifying that international large-scale assessments are leading to the standardization of education systems (Spring 2008), and the suggestion that these assessments are facilitating policy borrowing and the homogenization of education over time (Rutkowski and Rutkowski 2009). There is the potential, in future, that as more countries participate in surveys such as TIMSS and compare their outcomes on these assessments with those of other countries, curriculum reform will drive further convergence of curricula, leading to a stronger core international science curriculum. There is also the potential that new core science topics will emerge, based on global issues and challenges, or to better reflect advances in science or in areas of science with more economic potential, such as genetics and genomics.

Although this research suggests that it is possible to identify the basis of an international core science curriculum, there are a number of caveats which it is important to acknowledge. Firstly, the core curriculum identified here is defined on the basis of countries that have taken part in at least one of the TIMSS cycles considered in this report. Whilst the number of participating countries is high, the sample cannot be considered representative of all countries, with a higher number of economically developed countries participating in TIMSS and a relatively small number of countries from Africa and South America for example. As a result, we can only state that there is evidence of a core curriculum amongst countries that participate in TIMSS.

Secondly our analysis only considers science content as opposed to other aspects of the science curriculum such as the emphasis placed on science investigation, or the skills emphasized within the curriculum, such as scientific reasoning or problem-solving. Although we attempted to collect information on these aspects of the curriculum, due to inconsistencies between TIMSS cycles in the way this information was collected, it was not possible to make such comparisons. These features of the science curriculum are very important, but it is not possible within the limitations of this analysis to establish whether there are any international core standards with regard to the curriculum emphasis placed on science investigation or on scientific reasoning or problem-solving skills.

A further challenge and caveat is that our judgement of whether there is evidence to support the basis of an international core curriculum uses the TIMSS curriculum questionnaire responses. It has already been noted that the accuracy of these questionnaires and how well they reflect school curricula will vary between countries for several reasons. Firstly, in all countries there will be variation in the way the curriculum is interpreted and enacted in schools. Additionally, in some countries, particularly those with a federal structure, different curricula are likely to be in place in different states or provinces. As a result, a single curriculum questionnaire response for these countries is unlikely to accurately reflect the curriculum content of each state or province within the country.

The curriculum questionnaire provides information on whether a country includes a particular topic in its science curriculum at Grade 4 and Grade 8. Whilst this provides a useful summary of what is included in the curriculum, it is high-level information, and does not provide information about the level of scientific detail included for the topic. For example, the exact content of the Grade 8 topic on cell structure and function could vary considerably between countries. In some countries, the curriculum could simply specify the basic structure and function of animal and plant cells, whilst in others, the curriculum may require more detailed knowledge of cell organelles and the specialization of cells for different roles. Therefore, whilst we have identified common core topics taught in the majority of participating TIMSS countries, there could still be wide variation in particular topics between countries.

Finally, whilst we have identified topics that could be considered part of an international core science curriculum at Grades 4 and 8, this analysis has not taken into account how countries assess their science curriculum. Assessment is likely to play an important role as high-stakes assessment is likely to influence the science curriculum and what is taught in schools. Again, due to inconsistencies in the way the assessment questions in the TIMSS teacher and curriculum questionnaires have been phrased over multiple cycles, it has not been possible to compare assessment arrangements for science at Grades 4 and 8. Obtaining this information would be illuminating as it would identify whether there are common or core approaches to assessing the science curriculum at Grades 4 and 8, in addition to the core topics included within assessment.

References

Astiz, M., Wisemand, A., & Baker, D. (2002). Slouching towards decentralization: Consequences of globalization for curricular control in national education systems. *Comparative Education Review, 46*(1), 66–88.

Cogan, S., Wang, H., & Schmidt, W. (2001). Culturally specific patterns in the conceptualization of the school science curriculum: Insights from TIMSS. *Studies in Science Education, 36,* 105–133.

Kjaernsli, M., & Lie, S. (2008). Country profiles of scientific competencies in TIMSS 2003. *Education Research and Evaluation, 14,* 73–85.

Klieger, A. (2015). Between two science curricula: The influence of international surveys on the Israeli science curriculum. *The Curriculum Journal, 26*(3), 404–424.

Robitaille, R., Beaton, A., & Plomp, T. (Eds.). (2000). *The impact of TIMSS on the teaching and learning of mathematics and science.* Vancouver: Pacific Educational Press.

Rutkowski, L., & Rutkowski, D. (2009). Trends in TIMSS responses over time: Evidence of global forces in education? *Educational Research and Evaluation, 15*(2), 137–152.

Spring, J. (2008). Research on globalization and education. *Review of Educational Research, 78*(2), 330–363.

Zanini, N., & Benton, T. (2015). The roles of teaching styles and curriculum in mathematics achievement: Analysis of TIMSS 2011. *Research Matters, 20,* 35–44.

Chapter 6
Future Directions and Topics for Further Research into the Globalization of Science Curricula

Abstract The impacts of globalization on science education and curricula are of considerable interest internationally, not least in terms of preparing a nation's students for employment in a rapidly changing world. This study was not a measure of the total science curriculum for each country considered, but a measure of the similarity of their intended science curriculum to the Trends in Mathematics and Science Study (TIMSS) framework; further research into the effects on the science curricula of countries that have not participated in TIMSS or using data from other relevant large-scale assessments would add an additional dimension to understanding the globalization of science curricula. Research exploring the processes by which countries embark on science curriculum reform would be an important avenue for further work in order to gain a better understanding of why countries decide to make the changes they do.

Keyword Globalization · Science curriculum · Science education
Trends in Mathematics and Science Study (TIMSS)

Making use of the extensive IEA TIMSS datasets has enabled us to empirically address a number of important questions pertaining to the globalization of science curricula over a wide time range. The research complements and builds on previous research on globalization of curricula, which has tended to focus on mathematics rather than science.

However, globalization of science curricula is a broad topic, and we were unable to explore all areas of interest given our limited resources and time. As a consequence, we here identify important areas for additional research to build further on this work.

Firstly, we focused primarily on countries' intended science curricula, as opposed to the implemented or achieved curricula. Devising a complementary research program to explore the extent to which countries' implemented and achieved curricula may be converging would be the next logical step, and the TIMSS teacher questionnaire is an obvious source of relevant data.

© International Association for the Evaluation of Educational Achievement (IEA) 2018   89
O. Stacey et al., *The Globalization of Science Curricula*, IEA Research
for Education 3, https://doi.org/10.1007/978-3-319-71532-2_6

Secondly, our focus was on the TIMSS science topics included in countries' curricula, as opposed to the skills emphasized within curricula. There would be value in investigating how these aspects of curricula change in countries over time; for example, whether there is a trend for an increasing emphasis on science investigation and problem-solving skills, as opposed to pure content knowledge.

Thirdly, we compared countries' science curricula using the TIMSS science framework. Whilst this framework covers a broad range of science topics, it is not an exhaustive list. It is therefore possible that some countries' science curricula included topics that were not covered in the TIMSS science framework. One way to overcome this limitation would be to conduct a study that makes direct comparisons of countries' science curricula using the original curriculum documents (as opposed to the TIMSS science framework) as the basis for the comparison.

Our research identified a number of changes in countries' science curricula over time, particularly in terms of the science content included. However, we did not explore the causal mechanisms or drivers supporting the development of curriculum changes at an international level. Research exploring the processes by which countries embark on science curriculum reform would be an important avenue for further work, in order to gain a better understanding of why countries decide to make the changes they do.

We investigated globalization in science curricula using the 1999 TIMSS survey as a baseline for Grade 8 and the 2003 TIMSS survey as a baseline for Grade 4. As discussed in Chap. 2, Rutkowski and Rutkowski (2009) have suggested that investigating globalization in education and the curriculum using data from the 1990s onwards is starting too late, because the effects of globalization were apparent far earlier than this. Extending the study of globalization back further in time may detect changes and convergence in science curricula that our study may have missed, but depends upon the availability of reliable comparative data.

Here, we focused on countries that participated in TIMSS. It could be argued that countries participating in such studies have, to some extent, already bought into a global education agenda, and thus research that explores and compares the curricula of countries participating in TIMSS against the curricula of those that do not would provide an alternative perspective and add an additional dimension to the study of the globalization of science curricula. Furthermore, as many developing countries do not participate in TIMSS, analysis that makes use of available curriculum data from these countries would augment any study of the globalization of science curricula.

Finally, our research has not made reference to, or used data from, other international large-scale assessments of science such as the Programme for International Student Assessment (PISA). Given its influence and prominence, and the addition of new PISA assessments (such as the PISA-based test for schools and PISA for Development), research which considers globalization in science curricula within the context of PISA and its assessment frameworks, would also provide important complementary information.

Summary of the Key Findings

- The literature review identified a number of factors contributing to globalization in science education and science curricula. These included the growing emphasis on education as a mechanism for economic growth, the increase in information technology in education and the role of international large-scale assessments in facilitating comparisons between countries. Despite these factors, however, there were other forces which mediated the globalization of science curricula, such as local culture and local interpretation of a centralized curriculum.
- The analysis found strong evidence that there have been changes in intended science curricula over the last 20 years. For example, every country analyzed in the investigation had made changes, although the extent of the changes varied considerably. The analyses also identified that more changes were made to intended science curricula at Grade 4 than at Grade 8, therefore suggesting a more fluid science curriculum in the primary phase.
- The analyses suggest that, in some respects, science curricula are becoming increasingly similar over time. For example, there was a trend for more of the TIMSS science topics to be included in the majority (over 80%) of participating countries' curricula. This trend was observed at both grades, although was more pronounced at Grade 4.
- The cluster and discriminant analyses indicate strong convergence in the science curricula of countries participating at Grade 8. The number of countries clustering into Group 2 grew at the expense of the number clustering into Group 1. This suggests that, over time, changes in intended science curricula mean an increasing number of countries with similar profiles in terms of the topics they do and do not include in their science curricula.
- At Grade 4, the strength of evidence for science curricula becoming increasingly similar over time is less strong, with the results of the discriminant analysis suggesting that changes in countries' science curricula have led to only mild convergence in curricula.
- At both Grade 4 and Grade 8 two distinct groups of countries could be identified based on their intended science curricula. At Grade 4 one group tended to have a broad and balanced science curriculum covering a range of topics, whilst the other group had less emphasis on physical science topics and a greater focus on life sciences topics.
- Similarly, at Grade 8, one group included a greater number of TIMSS science topics in its intended curriculum than the other. However, the topics omitted from the group which included a more restricted number of science topics in their curriculum were not confined to one specific area of science, unlike Grade 4.

- Finally the analysis identified core science topics that were present in the intended curricula of the vast majority of countries included in this analysis. At Grade 4, these core topics tended to be concentrated in the life sciences. By contrast, at Grade 8, they were distributed more evenly across the science subjects.

Reference

Rutkowski, L., & Rutkowski, D. (2009). Trends in TIMSS responses over time: Evidence of global forces in education? *Educational Research and Evaluation, 15*(2), 137–152.

Appendix A:
Literature Review Search Strategy

Bibliographic Databases

The search strategy for each database reflects the differences in database structure and vocabulary. Throughout, the abbreviation "ft" denotes that a free-text search term was used and the symbol * denotes truncation.

Australian Education Index (Searched Via Proquest, http://www.proquest.com/, 13 October 2016)

#1	Curriculum development
#2	Globalisation
#3	Global approach
#4	Alignment (ft)
#5	Curricula alignment (ft)
#6	Curriculum alignment (ft)
#7	Curricular reform (ft)
#8	Curriculum reform (ft)
#9	Global alignment (ft)
#10	Globali*ation (ft)
#11	Internationali*ation (ft)
#12	Policy borrowing (ft)
#13	#1 or #2 or #3 ⋯ #12
#14	Primary school science
#15	Secondary school science
#16	Science curriculum
#17	Science education
#18	Elementary school science (ft)
#19	Science instruction (ft)
#20	#14 or #15 or #16 ⋯ #19
#21	#13 and #20

© International Association for the Evaluation of Educational Achievement (IEA) 2018
O. Stacey et al., *The Globalization of Science Curricula*, IEA Research
for Education 3, https://doi.org/10.1007/978-3-319-71532-2

#22   TIMSS (ft)
#23   Trends in international mathematics and science study (ft)
#24   #22 or #23
#25   #20 and #24.

British Education Index (BEI; Searched Via EBSCO, https://www.ebsco.com/, 12 October 2016)

#1   Curriculum alignment
#2   Curriculum planning
#3   Curriculum change
#4   Education and globalization
#5   Curricula alignment (ft)
#6   Curriculum development (ft)
#7   Curricular reform (ft)
#8   Curriculum reform (ft)
#9   Global alignment (ft)
#10   Global approach (ft)
#11   Globali*ation (ft)
#12   Internationali*ation (ft)
#13   Policy borrowing (ft)
#14   #1 or #2 or #3 ⋯ #13
#15   Science—Study and teaching
#16   Science—Study and teaching (Elementary)
#17   Science—Study and teaching (Secondary)
#18   STEAM education
#19   STEM education
#20   Elementary school science (ft)
#21   Secondary school science (ft)
#22   Science education (ft)
#23   Science curriculum (ft)
#24   Science instruction (ft)
#25   #15 or #16 or #17 ⋯ #24
#26   #14 and #25
#27   TIMSS (ft)
#28   Trends in international mathematics and science study (ft)
#29   #27 or #28
#30   #25 and #29.

The Education Resource Information Center (ERIC; Searched Via EBSCOhost, https://search.ebscohost.com/, 12 October 2016)

#1	Alignment (Education)
#2	Curricula alignment (ft)
#3	Curriculum alignment (ft)
#4	Curriculum development
#5	Curricular reform (ft)
#6	Curriculum reform (ft)
#7	Global alignment
#8	Global approach
#9	Globali*ation (ft)
#10	Internationali*ation (ft)
#11	Policy borrowing (ft)
#12	#1 or #2 or …. #11
#13	Elementary school science
#14	Science curriculum
#15	Science education
#16	Science instruction
#17	Secondary school science
#18	#13 or #14 or ⋯ #17
#19	#12 and #18
#20	TIMSS (ft)
#21	Trends in international science and mathematics study (ft)
#22	#20 or #21
#23	#18 and #22.

Journal "hand searches"
International Journal of Science Education (see http://www.tandfonline.com/loi/tsed20).

Website searches
International Association for the Evaluation of Educational Achievement (*IEA*; http://www.iea.nl).

Appendix B:
Questions Included in Selection Log

The following questions were included in the selection log used to appraise the relevance and suitability of articles for inclusion in the literature review.

(1) Who are the author(s) of the research?
(2) To which organization(s) are the author(s) affiliated?
(3) Which country was the research originally published in?
(4) What date was the research published?
(5) What type of publication (e.g. journal, book, government report) is the research published in?
(6) What is the main focus of the research?
(7) What age range does the research consider?
(8) Which country or countries does the research consider?
(9) Does the research provide evidence on the influence of globalization on school curricula?
(10) Does the research focus on science curricula? If not what subject(s) does it consider?
(11) Does the research make use of data from international large-scale assessments? If so, which ones TIMSS, PISA or others?
(12) Based on the responses to questions 1–11, an assessment of how relevant the research is to this study.

© International Association for the Evaluation of Educational Achievement (IEA) 2018
O. Stacey et al., *The Globalization of Science Curricula*, IEA Research
for Education 3, https://doi.org/10.1007/978-3-319-71532-2

Appendix C:
Example Review Template

Full reference
Aims of research
Information from source relevant for background/context
Country/countries included in study
RQ 1 What forces/factors are affecting/causing/influencing globalization in science education?
RQ 2 What evidence is there forthe globalization of science curricula over time?
RQ3 What approaches/methods have been used to investigate globalization of science curricula? (if appropriate) How have international data sets been used in the article?
Is this an empiricalstudy?
If yes is it a quantitative or qualitative study?

© International Association for the Evaluation of Educational Achievement (IEA) 2018
O. Stacey et al., *The Globalization of Science Curricula*, IEA Research
for Education 3, https://doi.org/10.1007/978-3-319-71532-2